METHODS FOR THE MEASUREMENT OF THE
PRIMARY PRODUCTION OF GRASSLAND

Methods for the Measurement of the Primary Production of Grassland

C MILNER & R ELFYN HUGHES

with contributions on

MEASUREMENT OF THE PRIMARY PRODUCTION OF DWARF SHRUB HEATHS

by C.H. Gimingham and G.R. Miller

and on

MEASUREMENT OF THE PRIMARY PRODUCTION OF ARID ZONE PLANT COMMUNITIES

by R.O. Slatyer

SECOND PRINTING

INTERNATIONAL BIOLOGICAL PROGRAMME

7 MARYLEBONE ROAD, LONDON NW1

BLACKWELL SCIENTIFIC PUBLICATIONS

OXFORD AND EDINBURGH

Printed in Great Britain by
BURGESS AND SON (ABINGDON) LTD.
ABINGDON, BERKSHIRE

Contents

Contents

Foreword

The overall purpose of the IBP is to prepare and carry out on a world scale a plan of research concerned with "the biological basis of productivity and human welfare". To make such a plan practical the programme is broken down into seven main sections, and the first of these deals with the productivity of terrestrial ecosystems (PT). Already one of the IBP handbooks has been concerned with this section, namely No. 2 *Methods for estimating the primary production of forests* by P.J. Newbould. This present number, in which methods for grassland, dwarf shrub heaths and arid zone plant communities are described, is a worthy addition to the series. It will be followed by other handbooks for Section PT now in preparation, one on secondary production by invertebrates and other small animals, and another on ecological and physiological methods for studying large herbivores.

Part I of this handbook, No. 6, contains a full review of the available methods for measuring the production of grassland. Parts II and III are more general in their approach and are intended to be used in conjunction with Part I, which describes many methods appropriate to the dwarf shrub heaths and arid zone plant communities.

The handbook is designed not with the object of standardising methods used at institutions which are already pursuing primary production studies, for many advances in the subject come from modification to and evolution of methodology; but it is designed for those individuals and institutions which require guidance on methods appropriate to the three community types described. At the same time use of the recommended methods provides some guarantee that results obtained all over the world will be comparable.

This volume, like others in the series, is to some extent provisional. The draft has been discussed with over thirty scientists currently involved in such studies, and many others have seen it without commenting in detail. Nevertheless improvements could still be made, and it is hoped that scientists using the handbook will correct or revise parts in which they have special knowledge and send their suggestions to the appropriate author. Indeed, it is hoped that revised and more definitive editions of many of the IBP handbooks will be called for before the conclusion of the programme in 1972, and that they will be useful to biologists for many years thereafter.

Dr C. Milner, who has served as editor of the book as a whole, as well as being principal author of Part I, was until lately a Senior Scientific Officer of the Nature Conservancy of Great Britain, based at the Bangor Research Station in Wales. He was head of the Productivity group, responsible to Dr R. Elfyn Hughes for the IBP programme in Wales. In May 1968 he moved to the Matador project, University of Saskatchewan, Saskatoon, where he has special responsibilities to Professor R.T. Coupland, its Director, for primary productivity measurement. Matador is the largest grassland research project at present operating under the IBP anywhere in the world. Dr Elfyn Hughes, specialist in the ecology and productivity of mountain grassland, is Director for Wales of the Nature Conservancy and head of the Bangor Research Station since 1960. He is a member of the IBP/UK/PT Sub-Committee and of its grassland working group.

Of the authors of Part II, Dr C.H. Gimingham is Reader in Botany at the University of Aberdeen and author of the British Ecological Society's "Biological flora of *Calluna*". Dr G. Miller is a Senior Scientific Officer in the Unit of Grouse and Moorland Ecology, a part of the Nature Conservancy in Scotland and of the University of Aberdeen. He is working on the productivity of *Calluna* which is browsed by grouse and sheep.

Dr. R.O. Slatyer, author of Part III, is Head of Environmental Biology in the CSIRO of Australia. His base is the Division of Land Research, Canberra.

E.B. WORTHINGTON
IBP Central Office
7 Marylebone Road
London, N.W.1

Acknowledgements

Part I is based on an original short draft produced by Dr R. Elfyn Hughes and circulated in 1965. An extended version was circulated in 1966 and the authors wish to thank the following for their helpful comments: Professor R.T. Coupland, Dept. of Plant Ecology, University of Saskatchewan, Saskatoon, Canada; Dr F.B. Golley, Institute of Ecology, Aiken, South Carolina, USA; Dr W.M. Johnson, Rocky Mountain Forest and Range Experimental Station, Laramie, Wyoming, USA; Dr J. Kvet, Academy of Sciences, Prague, Czechoslovakia; Dr H. Lieth, Botanisches Institut, Stuttgart, Hohenheim, Germany; Dr G.R. Miller, Nature Conservancy, Banchory, UK; Professor Milthorpe, University of Nottingham, UK; Professor P.J. Newbould, New University of Ulster, Coleraine, Northern Ireland; Professor M. Numata, Chiba University, Japan; Professor J.D. Ovington, Australian National University, Canberra, Australia; Dr R.O. Slatyer, CSIRO, Canberra, Australia; Dr R.G. Wiegert, University of Georgia, USA; and Dr T.E. Williams, Grassland Research Institute, Hurley, UK.

The complete handbook in corrected draft was discussed with members of the Grassland Working group meeting in Saskatoon, Canada in 1967. This included, in addition to several of the above: Dr L.C. Bliss, University of Illinois, USA; Dr F.C. Evans, University of Michigan, USA; Dr M.H. Gonzales, Centro Nacional de Investigacione, Chihuahua, Mexico; Dr M.J. Hadley, I B P/PT Section, Paris, France; Dr G.A. Petrides, Michigan State University, USA; Mr Ingvi Thorsteinsson, Agricultural Research Institute, Iceland; and Dr G.M. Van Dyne, Colorado State University, USA.

The authors of Part I would particularly like to thank Dr D.F. Westlake of the Freshwater Biological Association for the extensive comments and additional references he provided. We have also had the benefits of many discussions with our colleagues at the Bangor Research Station, in particular Dr D.F. Perkins and Professor A.M. Schulz on study leave from the University of California, Berkely and Mr J. Dale, Mr I. Rees and Mr

Ian G. Crook who provided initial drafts on which sections 3.1, 4.1 and 4.85 respectively are partially based. Although the suggestions made have been incorporated as far as possible in this version it has not always been practicable to reconcile the many different viewpoints and the authors and editor bear full responsibility for the form of the handbook, and for its omissions or mistakes.

Dr T. Pritchard helped in the editorial stages of the handbook and this assistance is gratefully acknowledged. The help of Mrs G. Crook in the drawing of the diagrams is also appreciated.

The editor would particularly like to thank Mrs A.G. Milner for much help with references and general correcting. The typists at the Bangor Research Station accepted the many drafts and difficult corrections with forbearance and this is gratefully acknowledged.

Part I

Methods for the Measurement of the Primary Production of Grassland

C. MILNER AND R. ELFYN HUGHES

1

Introduction

Grassland may be defined floristically as a plant community in which the Graminiae are dominants and trees absent. However, it is often more useful to consider it physiognomically or structurally as a plant community with a low growing plant cover of non-woody species. This definition includes therefore such communities as the early successional stages following the abandonment of arable land (old field) which contain a high proportion of dicotyledonous species and communities dominated by Cyperacae or Juncaceae, including those of arctic and alpine regions. It also includes grassland dominated by annual and ephemeral plant species occurring in arid regions, the main characteristics of which are dealt with in Part III of this handbook. Despite some structural affinities with grasslands, the dwarf shrub communities present unique problems and require methods specifically developed for use in such communities. One of this type of community, common in Britain and Scandinavia, is discussed in Part II of the handbook.

In view of the large area of the world occupied by grassland and its high potential for food production, the study of its primary production and the factors limiting this are of particular importance. A considerable volume of literature exists on the measurement of production in sown grasslands and rather less on the extensively grazed natural or sub-climax grasslands. This section of the handbook deals mainly with natural or semi-natural grasslands grazed by large herbivores (domestic or wild) or by smaller vertebrates such as rodents. Methods are also discussed for study of the much smaller acreage of grasslands which are not apparently utilised by vertebrates. Although many of the methods discussed are applicable to and derived from studies of intensively managed grasslands, in general the handbook is not principally concerned with this type of highly artificial plant community.

2

Terminology

IBP have issued a short list of definitions and symbols to be used in production studies (IBP News, No. 10, p. 6–8). Workers should ensure that they adequately define concepts used in their studies but should use the published list as a guide. The definitions below are those most likely to be used in practice and essential to the understanding of this handbook.

2.1 Ecosystem

In this handbook the term ecosystem will be used according to the classic definition of Tansley (1935), i.e. a functional unit consisting of organisms (including man) and the environmental variables of a specific area. This is a concept similar to the Russian "biogeocoenosis" with which, for practical purposes, it may be considered synonymous. The ecosystem concept is further discussed in Section 2.25.

2.2 Net Primary Production

2.21 Net primary production is the biomass or biocontent (total energy content) which is incorporated into a plant community during a specified time interval, less that respired. This is the quantity which is measured by harvest methods and which has also been called net assimilation or apparent photosynthesis. The net primary production can be qualified as in 2.23 below.

2.22 Net primary production can be expressed in mathematical symbols as shown below. This is the basic method quoted in IBP News 2.

B_1 Biomass (or biocontent) of a plant community at time t_1
B_2 Biomass (or biocontent) of the same plant community at time t_2

$\Delta B = B_2 - B_1$ Biomass change during the period t_1—t_2

L   Plant losses by death and shedding during t_1—t_2

G   Plant losses by consumer organisms, e.g. herbivorous animals etc. during t_1—t_2

Pn   Net production by the community during t_1—t_2

If ΔB, L and G are correctly estimated, Pn can be calculated as Pn = $\Delta B + L + G$. Providing always that the terms are adequately defined, it is perfectly acceptable to qualify net primary production with a descriptive noun, i.e. aerial, root, herbage, etc. For example:—

2.23 **Net primary aerial production** is the biomass or biocontent (total energy content) which is incorporated into the aerial parts (leaf, stem, seed and associated organs) of the plant community. This quantity is that which is usually measured by agronomists and is the ecosystem parameter of most value when large vertebrates are the principal herbivores.

2.24 **Gross primary production**—this is normally defined as the assimilation of organic matter or biocontent by a grassland community during a specified period; including the amount used up by plant respiration.

2.25 It is useful at this stage to consider the measurement of primary production within the conceptual framework of the ecosystem. The ecosystem is the fundamental unit of study in IBP and an important part of the IBP philosophy is that it is necessary and useful to study as many components of the ecosystem as possible. The relationships of the components are shown in Figure 1.

Although the measurement of the components shown is important to an ecosystem study, it will be apparent that measurement of many of these will allow an indirect measure of primary production or reduce the error of the estimate of net primary production.

Measurement of carbon dioxide uptake during photosynthesis and oxygen uptake during respiration allows the precise measurement of gross and net primary production. Although there are methodological problems and the interpretation of the observed values is sometimes difficult, the gaseous exchange method of measurement is the only one available for estimating gross primary production and respiratory loss. These methods are described in Sections 5.2 and 5.4.

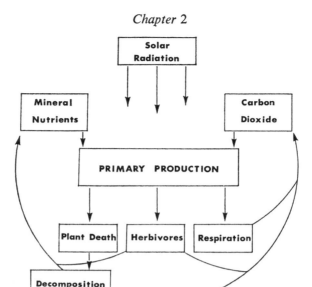

Figure 1. The relationship of the main components of a stylised grassland ecosystem. (Flow lines represent the paths of calories, organic matter or mineral nutrients as appropriate.)

Plant death and decomposition is a major source of error in the use of harvest methods for determining net primary production. This is dealt with in Section 4.72. In certain ecosystems it is possible indeed to calculate net primary production from a measurement of the standing crop of dead herbage and its instantaneous rate of decomposition. The removal of net primary production by herbivores is similarly a major source of error in the measurement of production when large grazing animals are present. In such ecosystems the harvest methods for determining net primary production require modification to allow an estimate of the ungrazed production. This also enables an imprecise estimate to be made of the amount of herbage entering the grazing animal population.

Although this handbook is concerned therefore with the primary producers in the ecosystem, the role of this trophic level cannot be conceptually separated from the others. It must be regarded as one variable of the complex functional unit defined as the ecosystem. The ecosystem, at whatever level of abstraction chosen, is the only valid study unit in IBP and this handbook dealing with one trophic level only should be read with this in mind.

3

The Study Area and Site Selection

3.1 Site selection

If the biome possesses several major grassland communities, study areas will be required in each community type. When finances and manpower are limited, choosing the communities of greatest study value often presents serious difficulties. This must be a local decision, but acreage, economic value, academic interest and ease of study (ease of access, structural simplicity, etc.) are important considerations. It may also be necessary to consider several community types as comprising an ecosystem, particularly if large herbivores move freely between them and this should be considered in the design of the study areas.

In each region it is useful and important to compare production of the natural climax grassland or extensive subclimax grasslands with sown swards subject to high input of management and plant nutrients. The grasses sown will be dependent on the region and should be those agronomically suitable for the environmental conditions. Nutrient input should be that normally applied to intensively managed grasslands in the region. The sown grassland study will require a similar layout to that shown in Figure 2 with the artificial community covering the entire area.

3.2 Site layout

The research area should have a sufficient area for the experimental and observational records envisaged and obviously this will vary with the grassland type and complexity of the programme. Newbould (1967) has suggested a scheme for woodland work, and a similar layout is useful for grassland studies as shown in Figure 2. The dimensions given are little more than suggestions for the immediate requirements of primary productivity studies. In every programme it would be desirable to have larger areas available for study. If animal experiments are to be undertaken, a much larger area will be necessary.

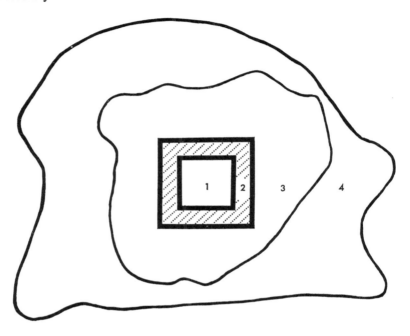

Figure 2.   The study site: 1, sample area: 2, buffer area: 3, measurement area: 4, study area (from Newbould, 1967).

The various parts of the diagram and their function are described below:

1   Sample area. This is used only for non-destructive measurements such as microclimatic recording, floristic studies or gas exchange photosynthesis measurements. It may be possible to utilise the area after the investigation has been functional for some time and sufficient non-destructive records have been taken. (Area required 0.1—1 ha.)

2   Buffer area. This should be an area of at least 10 m. width around the sample area. It should not be subjected to any disturbance which could affect the sample area.

3 Measurement area. This is the area in which vegetation and roots are harvested and in which soil samples are taken. In most grassland communities sampling in this area should not seriously affect the sample or buffer areas. (The area required will vary between 1—20 ha depending on circumstances.)

4 Study area. This serves mainly as a large scale buffer area and should be controlled by the research worker concerned. In many programmes it will be part of the research station and used for the study of wild or tame herbivores and for the maintenance of study animals. (Area required 20—200 ha.)

The areas need not necessarily be arranged as shown although obviously the buffer zones must surround the sample and measurement areas. In some grassland studies even the measurement area is not *necessarily* destroyed by the cutting of vegetation or other sampling techniques as the micro-climate of the area is unlikely to be seriously affected by scattered quadrats cut within it. However, sampling for root biomass in certain soil conditions may result in destruction of the measurement area by alteration of normal drainage patterns. The possibility also exists that trampling by research workers will modify both the floristic composition and productivity of the grassland community. This should be minimised by the provision of marked pathways.

3.3 Site description

An accurate and precise description of the study area and its environment is of great importance and a preliminary survey will be required before the commencement of the study. It is essential to be able to assign the area of study to a broad phytogeographical group and desirable to assign it more precisely to a phytosociological group. Methods of surveying, describing or classifying vegetation are not discussed here, but the following reference works should be consulted: Brown (1954), Cain and Castro (1959), Kershaw (1964), Curtis (1959), Joint committee on range research methods (1962).

If possible, information on soil type and geological features of the entire area should be obtained in the early stages and may affect the choice of the study site.

Climate should be measured as appropriate and, in the absence of specialist advice, observations on light, temperature, precipitation, evaporation, and wind should be made using recommended methods. The establishment of meteorological stations should be afforded high priority to allow characterisation of the local climate as soon as possible in the investigation.

4

Measurement of Net Primary Production
by Harvest Methods

4.1 Introduction

The measurement of net primary production by measuring the biomass or biocontent of the plant community at the beginning of a study period and again at the end, allowing the calculation of increment by subtraction has been widely used and recommended. In grassland studies this has mainly been used to determine net primary aerial production but can be applied, in theory at least, to the complete plant community.

There are two methods which have been used by agronomists on intensively managed grassland, only one of which will be considered in detail in the handbook. The method of pretrimming the grass sward and measuring regrowth is rarely applicable in ecological studies despite its apparently higher accuracy (Boyd, 1949). In many grassland situations pretrimming is a very drastic treatment which is likely to modify subsequent growth to such an extent that measurement is valueless. The method is not therefore dealt with in this booklet. The difference method which has normally been used by ecologists should be used.

4.2 Sample plot size and shape

The choice of sample plot size and shape to be harvested in grassland studies is limited by two main features. These are the necessity to obtain an acceptable level of accuracy in determining the standing crop of the study area and the practicality of harvesting the sample required.

4.21. The time required to clip the sample plots is the limiting factor in determining the number clipped and hence the accuracy of the estimate. Consequently, a compromise must be made between accuracy and the time required for field sampling, i.e. the cost of the sampling. It is necessary in all stations therefore to determine as far as possible the optimal number, shape and size of sample plots by a preliminary trial before commencing the main study. Normally, the outer region of the study area (Area 4, Figure 2) would be the appropriate place for this.

4.22 It has been traditional in grassland studies to utilise square sample plots (quadrats), but there is considerable recent evidence that this shape is not the most appropriate for maximum accuracy. In general, the variance of the samples should be at a minimum in non-randomly distributed herbage standing crop (its usual condition) when the sample unit is a rectangle. Similarly, variance should be at a minimum with small sample plot sizes and larger numbers of sample plots. However, it must also be remembered that if the number of samples are fixed for some reason, greater accuracy will be obtained by increasing sample plot size. It is, however, necessary to strike a balance between, for example, the possibly lower variance of small plots and the increased edge effects which will occur. Coleman (1959) and Van Dyne *et al.* (1963) have more recently shown the considerable advantages of circular sample plots and it seems possible that this shape is the most suitable for most grassland types although not yet tested on all possible vegetation types. Examples of the range of sample plot sizes is given in Table 1 below.

Table 1

Vegetation Type	Sample Plot Size (converted to metric units)	Authority
Alpine	2 dm. × 5 dm. (4 per site)	Bliss, 1966
Range	890 cms.^{2}	Campbell and Cassady, 1949
Range	53 cms. radius (circular)	Coleman, 1959
Mountain Grassland	50 cms. × 50 cms.	Milner and Perkins (unpub.)
Sagebrush/ Grass Range	2322—2787 cms.^{2}	Pechanec and Stewart, 1940
Alpine	23.5 cm.^{2} (100 per site)	Scott and Billings, 1964
Natural Grassland	50 cms. × 50 cms.	Shimada, 1959
Range	185.8 cm.^{2} (circular)	Van Dyne *et al.*, 1963
Hill Bunchgrass	30.5 × 122 cms.	Van Dyne *et al.*, 1963
Old Field	0.187 m.^{2}	Wiegert, 1962
Prairie	1 sq. m.^{2}	Many authorities

The decision on sample plot size and shape therefore must be taken following a trial in which a selection of the usual small plot sizes and shapes for the grassland type are tested. It is important also to ensure that an adequate number of plant units are included within the sample plot irrespective of other considerations.

4.3   Number of sample plots

4.31 The total number of sample plots depends on the degree of precision required which must, however, be matched against economic considerations, i.e. the time required to cut the sample plots. For many purposes, an error of 10% of the mean is an acceptable standard and should be approached in IBP studies. In order therefore to decide the number of sample plots required with a given sample plot size and shape, a small number of samples are cut. The number will vary with the subjective judgment of variability, but should not be less than ten. The number of plots required (N) is then approximately given by

$$N = \left(\frac{ts}{D\bar{x}} \right)^2 \quad *$$

where s = standard error of trial plots
D = required level of accuracy, expressed as a decimal (i.e. 0·1 in this case)
t = obtained from standard statistical tables.
(This formula applies only to continuous data.)

4.32 A graphical method of deciding when an acceptable number of samples has been *obtained* has been described by Greig-Smith (1964). The use of this is illustrated in Figure 3 and has the advantage of indicating whether further sampling is necessary. The mean of the first five, ten, fifteen, twenty, etc. observations is calculated and plotted against the number of observations. It will be seen that the mean at first fluctuates, steadying as the required number of samples is reached. In the example given the optimum number of samples (quadrats) is 25—30. Despite the subjectivity of this approach, it is a useful technique in many situations.

4.33 In the techniques described it has been assumed that an arbitrary sample plot shape and size has been chosen and economic considerations

---

* A formula which takes into account that s itself is an estimate is given by Cochran & Cox (1957).

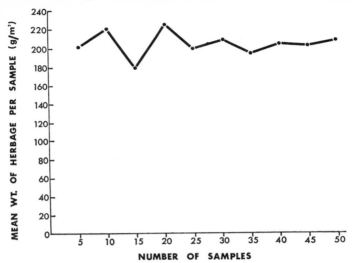

Figure 3. Subjective estimate of sampling accuracy (Greig Smith, 1964). Mean of first 5, 20, 15, . . . samples. Dry wt./m.² (Data from Milner & Perkins, unpublished.)

have been ignored. Van Dyne *et al.* (1963) and Wiegert (1962) have both investigated the influence of sample plot size and shape on the economic efficiency of sampling and should be consulted for the detailed methodology. In designing a trial of this type, however, certain basic considerations should be taken into account. The various sizes of quadrat should not be obtained by bulking contiguous or nested plots owing to the considerable edge effects which can cause misinterpretation of the results (although Wiegert (1962) uses nested quadrats, Van Dyne *et al.* (1963) show this causes bias in his data). The various shapes and sizes should therefore be located at random over the area selected. If this is not done, it is inevitable that the yields of the various sample plots will be determined by the previous plots cut as indicated in Van Dyne *et al.* (1963). To reduce variability in cutting technique, it is necessary to cut all plots using the same operator and clipping equipment. In analysing the effectiveness of each plot size and shape, it is necessary to take into account not only efficiency, i.e. the lowest variance for a given time cost, but also the production of normally distributed data rather than skewed. (Transformation of the data may, however, take care of this difficulty.) It is also important to test whether there is heterogeneity of variance among the various plot sizes since most common statistical tests are invalid if this is so. Van Dyne *et al.* (1963) give the appropriate test.

Although it is important to determine optimum quadrat size and shape if possible, it should be noted here that even a subjectively chosen size, shape and number will allow an estimate whose standard error can be determined and quoted. It is also important to realise that IBP is of limited duration and in most studies it is vital to obtain net primary production data at the earliest opportunity. It is also probable in most cases that the measurements of secondary production will be at a lower order of accuracy than those described in this handbook.

If at all possible, however, even a very limited preliminary sampling programme should be attempted.

4.34 The sampling units should be randomly located within the vegetation types studied using standard statistical methods. Van Dyne (1960) has suggested the use of aerial photographs for random location of sample units and his technique is appropriate for many studies. Where stratification of samples is necessary, Van Dyne's method would be appropriate, as would other systems for ensuring the random location of sample plots within the stratification groups, e.g. Reppert *et al.* (1962).

4.4   Harvesting techniques

4.41 All harvest methods for above ground vegetation depend on clipping a measured area of vegetation uniformly and at a predetermined height above the surface of the ground. In grasslands this is not a difficult technical problem (cf. woodlands) but still requires careful consideration and testing in each case. In particular, it is important to specify the cutting height and consideration must be given to a number of factors affecting this: the two most important being (1) the species constituting the sward and (2) the height/weight relationship of the species. The surface of the ground with its associated litter would normally be the fixed point, but in certain situations with a bryophyte layer, for example, this point may be difficult to decide and cause errors in short grass communities with low biomass. Individual operators, unless given specific instructions, show consistent differences in the height of vegetation left after cutting (Milner and Perkins: unpublished data) and this should be taken into account. In every case some estimate of the weight of vegetation left after cutting should be obtained. This can be done using coring techniques in conjunction with root measurements (Section 4.6).

4.42   The actual cutting device depends on the type of vegetation cover to be sampled, but hand shears are widely used. These can be hedge clippers for two-handed operation in tall grass cover (e.g. many tropical grassland communities) or shears for single-handed operation such as sheep shears or their garden equivalent. Single-handed shears are available in which the cutting blades and the operating handles are at a 90° angle allowing clearance between the hand and the ground surface. These are particularly valuable in short grass swards.

4.43   Clipping herbage to determine yield is a costly part of any production study. Many mechanical devices have been used to reduce this to an acceptable level. The main types of mechanical clippers are discussed below.

A self-propelled, petrol-driven, reciprocating blade cutter with one — two feet cutting width has been used for large plots and could be useful in tall grass situations. For smaller samples, however, it is not sufficiently close-cutting. Hand-held, reciprocating blade hedge clippers have been described, either powered by direct current from twelve-volt batteries, by alternating current from petrol generators, or cordless with an integral rechargeable battery. (For example, Matches, 1963.) Such clippers often require modification for use on grasses and these are usually available from the manufacturers. A rotary lawn mower has been used by McGinnies (1959) for sampling range herbage and may be useful in certain conditions.

None of the above are, however, particularly suitable for grass under about 10 cms tall, and hand methods have often been found essential for such vegetation. More recently, methods utilising powered sheep clippers have been evolved. Alder and Richards (1962) have described the use of such a three-inch sheep shearing head powered in several ways, all proving satisfactory. This equipment was particularly valuable for shorter grass species, but even better are the vacuum clippers described by Becker (1959) and Van Dyne (1966) for use on short-grass rangeland dominated by low growing species such as blue grama (*Bouteloua gracilis*). The equipment basically consists of a sheep-shearing head combined with a vacuum cleaner which collects the cut material and reduces the losses of plant material occurring when swards of less than 5 cms are cut by conventional methods. A certain amount of litter and soil is picked up by the machine but Van Dyne gives appropriate regressions of net weight (i.e. weight of herbage) on gross weight and discusses the use of separation techniques to calculate net weight. The vacuum clipper certainly increases the speed of sampling by five to ten times and even including the time required for separation to obtain net weight (20—40% of samples) should provide a worthwhile saving of time. Botanical separation is difficult when this technique is used, but the method can be applied where separation of constituents is required if the laboratory point analysis procedure (Heady and Van Dyne, 1964) can be used. (See Section 4.51.)

4.44   The placing of a complete quadrat or circle of metal or other material in dense grassland vegetation as a cutting guide is often a difficult and not completely objective operation. The insertion of the quadrat into the base of the community in dense or lax grassland may cause difficulties, which may be partially overcome by using very thin material to manufacture the quadrat or by using the rigid, open-ended quadrat described by Thilenius, (1966). This can be pushed into the tangled base of the grassland allowing efficient separation of the leaves within the sampling area from those outside by then raising the quadrat through the herbage. Westlake (pers. comm.) also uses a quadrat which can be inserted into the base of the vegetation. Such methods are not easily adopted if a circular quadrat is used.

   It is possible that edge effects will be important in species with lax aerial parts and techniques for harvesting in this type of sward should be carefully standardised.

4.45   The cut material from conventional cutting techniques may require either sub-sampling in the field (after determining total fresh weight) or transportation in bulk to the laboratory. In the former case the technique proposed by Hilpoltsteiner (1960) appears promising. This utilises a portable chaff cutter which can quickly cut two-to-three kilograms of herbage into five mm lengths for sub-sampling. This would be of particular value when the herbage standing crop is high and transportation difficult.

4.46   Polythene bags should be used for the storage of fresh cut herbage, if possible, and Edwards (1965) has recently shown that samples can be kept in such bags for several hours with only small losses of moisture and no significant loss of dry matter. It seems likely, however, that considerable changes in the proportions of the various organic constituents (e.g. soluble carbohydrates) would result from this treatment, and it is always desirable to dry or deep freeze samples as soon as possible.

4.5 Botanical separation and associated measurements

4.51 Measurement of species composition. Separation of the harvested material into species or into leaf, stem, and flower is a necessary task in any study other than the very simplest. This separation is, however, particularly time-consuming and costly and methods should be devised, if possible, to reduce this cost to a minimum. Hand separation requires least skill and is probably the most accurate of the available techniques. However, sampling problems are often acute and care must be taken to ensure satisfactory sub-sampling. "Production line" techniques with unskilled labour are often the best compromise between accuracy and speed, and if species are morphologically distinct, may be perfectly adequate.

The point quadrat techniques of Wilson (1960, 1963, 1965 *a* and *b*) (see Section 4.55) with suitable calibration against cut and separated samples, might provide a satisfactory measure of the relative weights of the various plant species in the sampling area. It has the advantage of providing relevent information on leaf area at the same time.

The laboratory point method of Heady and Van Dyne (1964) has been successfully used on a variety of samples and is particularly useful when the cut vegetation is short or has been chopped into short lengths for sampling. The technique, which utilises a binocular microscope and movable tray in which the sample is spread, can be quickly standardised and in communities with species which are separable under a binocular microscope, provides a very useful measure of the relative proportions of the species. It does not, however, provide samples for chemical analysis which must be obtained by separation.

4.52 Measurement of dead/green ratio. Wiegert and Evans (1964) have shown that a simple measurement of herbage standing crop at the beginning and end of a study period (which may be repeated several times during the growing season) is not sufficient for a measurement of net primary production. Perennial grasses produce tillers which die or flower throughout the season and consequently a simple harvest technique based on green plant material will underestimate net primary production by a variable, but considerable amount. Brougham (1962) has discussed this with particular reference to clover swards and should be consulted for a review of the problem. Westlake, D. F. (pers. comm.) has also recognized the problem of tiller death and has used an interesting approach based on fish growth techniques using graphs of the number of shoots surviving out of the number originally initiated in a month against mean weight at monthly intervals. The area under a smoothed curve represents the production of each cohort (i.e. each set of shoots initiated each month). This method will be outlined in the PF manual and should be consulted for details. A similar method has been mathematically examined by Skellam (1967) with particular reference to animal populations. Perkins, D. F. (pers. comm.) has also considered this problem in oceanic mountain grassland species in North Wales.

Allowance must be made therefore for changes in dead material from one harvest to the next, and a measure of dead material present at any harvest is necessary. Hand separation of the fresh herbage has the obvious merit of objectiveness and the provision of samples for subsequent analysis but the grave disadvantage of high labour cost.

4.53 Two techniques have been suggested to reduce this cost, that of Hunter and Grant (1961) and Heady and Van Dyne (1965). Hunter and Grant used an extraction method for determining green dry matter which involved the extraction of dried ground plant material with methanol and subsequent measurement in a photoelectric absorbtiometer. The weight of green is calculated from a previously determined calibration curve. This useful technique offers no opportunity for analysis of the green or dead plant material but has many practical advantages. Heady and Van Dyne used a visual "point" method in the laboratory for species separation (see Section 4.51) which could be simply adapted to dead/green separation.

4.54 Measurement of leaf area. In many investigations, particularly those of an advanced nature, the determination of total leaf area is a useful additional measurement. This information, although valuable, is laborious to collect and subject to considerable error. For this reason many attempts have been made to devise instrumental methods for area measurement of detached leaves, most of them unfortunately of little value in grasslands.

Instruments in which the leaf is placed between a light source (often red) and a photoelectric cell have been commonly used and Donovan *et al.* (1958) and Pilet and Meylan (1958) give details. Other instruments, such as that of Orchard (1958 and 1961) which utilises a scanning device and others using cathode-ray tubes, have to some extent superseded photoelectric methods. The method of Jenkins (1959) ulilises the drop in air flow when leaves are introduced into a constant airflow and this provides a useful and quick determination with many advantages. The method of Jones (1961), which consists of placing leaves on a stiff screen and relating the rate of passage of a standard volume of sand to the area, appears accurate but unnecessarily complex.

All these instrumental methods are of limited value in grasslands, particularly when fine-leaved grasses are present in the sample and the method of Kemp (1960) is the most useful in grassland studies. The area of a leaf is given by the equation:

$$A = kLB$$

where L = leaf length
B = breadth at midpoint
A = area
and k = a constant determined for the species under investigation.

The method has the considerable advantage that leaf sheaths and cylindrical leaved species are capable of measurement by this method, both structures being treated as cylinders. Length and breadth can be measured using squared paper or a microscope with a graticule, depending on size of species.

4.55 The inclined point quadrat developed by Wilson (1960) for the determination of leaf area in intact communities is of considerable interest and has considerable value in grassland studies. The methods and mathematics, however, have been developed for temperate broad leaved grasses and may require further elaboration in other grassland types. Considerable detail on both methodology and theory is provided in Wilson (1963, 1965 *a* and *b*) and Philip (1965) and these accounts should be consulted by interested workers. The inclined point quadrat technique has a particular value if used in conjunction with a non-destructive measurement such as described in Section 5.5.

4.6 Measurement of root production

4.61 The root system of grasslands present many problems to the production ecologist. They constitute a large and important proportion of the net primary production and standing crop in grassland communities but present formidable sampling problems. These are associated with two main features of the root system of grasslands, the fact that the roots in many soil types are inextricably attached to soil particles and also that there are no easy methods for separating living and dead roots. For these reasons, few completely successful techniques have been evolved for use in ecological investigations. The studies of Bray (1963) and Dahlman and Kucera (1965) are among the few which have been concerned with the dynamics of root production and its effect on estimates of net primary production.

4.62 Considerably more investigations have been carried out on the root systems of sown grassland and Troughton (1957, 1958, 1959, 1960, 1961) gives useful data on root growth which, although mainly concerned with cultivated grasses, may nevertheless provide valuable methodology for the laboratory or greenhouse study of non-cultivated species.

If a laboratory or greenhouse study is considered useful, the methods of Muzik and Whitworth (1962) and Lavin (1961) are useful. Both methods are of use in studying the growth and behaviour of individual roots and utilise a glass-sided box against which the roots of seedlings grow. Translating observational data of this type into quantitative terms is, however, difficult and of little direct relevance to intact grassland communities. Studies of spaced plants in the field, although slightly more relevant, have little real value but may be of use in an initial approach. Roder (1959) has used this technique with cultivated species.

4.63 Measurement of the root production of grassland requires some initial semi-quantitative descriptive data on the root systems involved. Considerable data of this type are available, with the studies of Weaver (1958 *a* and *b*), Dahlman and Kucera (1965) and Coupland and Johnson (1965) being of particular significance. Such studies are necessary to determine the extent and type of root system in order to devise sampling techniques for more quantitative studies. All the studies involve the removal from the sward of a complete soil core or block with roots. The root systems are then washed free of soil for examination in the laboratory. These studies show the considerable depths to which the root systems of grassland penetrate and jndicate something of the problems involved in obtaining a representative sample.

4.64 It is possible to calculate root productivity from a combination of observations on length of life of roots, turnover time and maximum standing crop, the appropriate relationship being

$$\text{root production} = \frac{\text{maximum standing crop}}{\text{turnover time}}$$

Maximum standing crop can be determined by one of the many variations of washing roots from a known volume of soil. Typical variations are described by Upchurch (1951), Fribourg (1953), Williams and Baker (1957), Deffontaines (1964), Il'in (1961), and Carlson (1965). Other methods which have been used are the flotation method of McKell *et al.* (1961) and the method of Dahlman and Kucera (1965), who utilised a 1% solution of sodium hexametaphosphate for preliminary soaking, followed by mechanical agitation in a 0·8% sodium hypochlorite solution.

4.65 However, all the methods described have the fundamental difficulty of separating living and dead roots and of removing soil particles adhering to root hairs. Colour and general form of the roots have been used to separate living from dead and with care and experience of the species under study, may provide an acceptable technique. Tetrazolium (2, 3, 5-triphenyl tetrazolium bromide) has been used to indicate viable root tissue more precisely (Jacques and Schwass, 1956). Unfortunately, this does not appear to be completely suitable although modification of this technique or the use of some alternative staining method may be of possible value.

The presence of soil in the final sample is inevitable but the degree of contamination may be determined by ashing the samples. If the soil is of mineral origin, this indicates the weight of roots in the sample (Willard and McClure (1932) and Williams and Baker (1957)). The ashing method is unlikely to be of much value in the determination of root weight in soils high in organic matter in various stages of decomposition.

4.66 Determination of turnover time is of considerable complexity and can often only be determined by regular sampling and separation over a year. Dahlman and Kucera (1965) suggest four times, based on previous phenological observations, namely

(a) Midsummer (high vegetative production)

(b) End of growing season (vegetative die back)

(c) Midwinter (no vegetative growth)

(d) Spring (immediately prior to growth).

It will, however, be obvious that such a sampling frequency is difficult to sustain and, in fact, in this case root production can be determined directly by difference methods. However, turnover time, which must be related in an undetermined fashion to the age of the roots, is a useful concept to bear in mind and may be useful as more efficient methods for measurement are developed (Tatarinova (1961)), Tatarinova used radio isotopes to study the physiological connections between shoots and roots and was able to make deductions about root longevity. This type of approach in study plants may allow extrapolation to the field condition.

Other authorities have utilised radio isotopes rather differently for root studies (mainly using P^{32}) which allow the calculation of root growth rate, root competition and the separation of root systems of individual plants. (Neilson (1964), Hall *et al.* (1953), Deklit and Talsma (1952)).

4.67 A useful technique of some value has been suggested by Troughton (pers. comm.) and used successfully on mountain grasslands in Britain (Milner and Perkins, unpub.). This consists of removal of a core from the sward to appropriate depth and replacing with soil containing no roots. After a given period, the identical core is again sampled and roots which have grown into the soil weighed. The technique is only possible in grassland where most of the roots are in the top few inches, but its simplicity is of considerable value and deserves further use and study.

4.7 Special problems of ungrazed grasslands

Although most grasslands are grazed by large herbivores, there are relatively small areas in all the major climatic zones which are not. These, however, may contain high populations of invertebrate herbivores. The problem of measuring net primary production in these grasslands are considerably reduced as no elaborate caging to exclude vertebrates is required, however they do have some particular methodological problems.

4.71 The invertebrate component of grasslands can be important (for example, see Wiegert, 1967) and the removal of primary production by this means is one of the errors involved in the harvest method in ungrazed ecosystems, although perhaps not an important one. In order to correct net primary production for the energy removed by invertebrates, a full study of the dynamics of the populations, their metabolism and intake would ideally be required. This is, of course, often the object of an investigation of secondary productivity and should be rigorously pursued in any IBP grassland ecosystem study (Section 2.25). It is difficult or impossible to measure the invertebrate consumers by a difference method owing to the difficulties of excluding invertebrates mechanically. However, methods have been used in woodlands (Newbould, 1967) involving the estimation of the area of leaf removed from a sample by invertebrates. The area removed is estimated by planimeter working on a photoprint of the leaves. This technique could be of value in grasslands with large grass species or many broad leaved species. Westlake (pers. comm.) has utilised the difference in weight between large numbers of matched pairs of grazed and ungrazed leaves of *Glyceria*. This gives an acceptable estimate in the case of large leaved grasses, but is difficult with fine leaved species. It is likely, however, that the best approach will be a detailed study of the invertebrate component, and if this is possible, the account of the IBP Warsaw working meeting on Secondary production (Petrusewicz ed. 1967) or Southwood (1966), should be consulted.

In many programmes, however, it will be impossible to obtain precise measurement of the invertebrates in the ecosystem and the consequent loss of net primary production must be accepted. It should, however, be possible to survey the major invertebrate groups in a semi-quantitative fashion.

4.72 Most ungrazed grassland communities will contain large amounts of dead material at certain times of the year which may well contribute structurally to the functioning of the ecosystem. Methods of describing and elucidating this function are not yet obviously available, although Wiegert and Evans (1964) have discussed dead material and its use in determining production (see Sections 2.25 and 4.52). Their method, which is extremely useful and should be used whenever possible, utilises the rate of disappearance of dead material combined with data on green and dead standing crop to calculate net primary production of the above ground plant parts. As discussed in Section 2.25, this data is important to the study of the ecosystem apart from its value in determining net primary production. The method described allows an estimate of mortality which clearly has an important effect on the accuracy of measurement of net primary production. The appropriate equations are given below:

(Standing crop data are expressed in grams/m² and instantaneous rates in mg./g. per day.)

Let t_1 = time interval in days

a_{i-1} = standing crop dead material at start

a_i = standing crop dead material at end

b_{i-1} = standing crop green material at start

b_i = standing crop green material at end

r_i = instantaneous daily rate of disappearance of dead material during interval

Let x_i = amount of dead material disappearing during an interval

$$(1) \qquad x_i = [(a_i + a_{i-1})/2]r_i t_i$$

Changes in standing crop of green and dead respectively are

$$(2) \qquad \Delta b_i = b_i - b_{i-1}$$
$$(3) \qquad \Delta a_i = a_i - a_{i-1}$$

Since Δa_i is the change in dead standing crop $(x_i + \Delta a_i)$ is the amount of material added to the dead standing crop (symbolized here by d_i).

$$(4) \qquad d_i = x_i + \Delta a_i$$

The growth during t_i is then given by

$$(5) \qquad y_i = \Delta b_i + d_i$$

where y_i is in g./unit area.

Equation (4) must be $\geqslant 0$, negative values indicating an error in one or more of the measured parameters.

It is possible, by using the techniques described by Wiegert and Evans, to calculate annual production given only the standing crop of dead and its instantaneous rate of decomposition. This may be of value in certain ecosystem studies.

4.8   Special problems of grazed grasslands

Grassland which is grazed by large herbivores is the commonest and economically most important of the life form and, in this situation, the measurement of net primary production presents two inter-related and mutually interfering problems.

4.81   The harvest method requires the measurement of standing crop at time intervals and in grazed situations this involves the protection of the accumulated growth from the grazing animal, to prevent an underestimate of net primary production. The exclusion of the grazing animal, however, imposes a different set of environmental conditions on the community immediately. This will have a considerable effect on the net primary production. Nevertheless, since it is obviously necessary to exclude large grazing animals, this effect must be accepted. The effect of exclusion can be minimised if the period of exclusion is kept low compared to the life cycle of the main species (see for example Green, 1949). In practice, however, sufficient growth must have taken place to give a measurable change in herbage biomass. This period will vary considerably with environmental or floristic differences and must be decided locally.

4.82 Cages for the exclusion of grazing vertebrates have been widely described and many variations are possible. The variations will be affected by the materials available locally, the labour available for moving, the size of sampling unit and most important, the animals to be excluded. It is unnecessary to give all the many variations which have been described, as the choice will be essentially a local one. However, to illustrate something of the range of variation involved, two cages are described. Smith and Sheets (1960) describe a very effective cage consisting of a cylinder of 11 gauge 2 in. $\times$ 4 in. welded mesh 4·55 ft. in diameter. Seven of these cages can be made when and where required, from a 100-ft. roll of the welded mesh transported to the site. This cage would be suitable for the exclusion of sheep or small ruminants of similar size. Many cages, particularly those designed for the exclusion of cattle or larger ruminants have a pyramidal form or inwardly sloping ends. The San Joaquin cage (Westfall and Duncan, 1961) is of this form and utilises a wooden frame.

Small ruminants (e.g. sheep) can be excluded by a very simple cage of non-rigid form with an open top providing the corners are guy wired to pegs. Welded cages may require guys also, but if such cages are square in form, unless very heavy, they may be less useful than the simple non-rigid cage owing to the habit of many ruminants of rubbing against a solid object. It is necessary to select the simplest and lightest design for the animals to be excluded and it is surprising in practice how light and simple this may be.

The cages described can be designed to exclude all sizes of herbivore dependent on the mesh size used. If burrowing animals are present, however, the cages will require burial to prevent the animals tunnelling into the cage. If very large mammals, e.g. elephant, occur in the experimental area, other methods of exclosure will be required and the use of electrified fencing may provide a suitable technique.

Galvanised wire cages can give rise to zinc toxicity or contamination of the herbage. This can be prevented by the use of a modern flexible resin paint applied to the cage after construction. The best method of application is by dipping which ensures complete coating of the structure. The labour involved will be very worthwhile, particularly as cage life will be extended very considerably.

Detailed descriptions of the range of cages which have been used will be found in the literature for further reading at the end of the handbook.

4.83 The effects of cages are apparently variable and associated with a changed microenvironment within the cage and protection from the possibly damaging effects of defoliation and trampling. This seems to outweigh the possible effect of removal of the stimulus of light grazing. The effect of cages is to increase the dry matter production by between 11% (Cowlishaw, 1955) and 15% (Jagtenberg and De Boer, 1957). However, in temperate latitudes, the effect is considerably reduced when rainfall is high. This is to be expected as the main effect of cages is to reduce wind velocity, to increase humidity, and to lower the transpiration rate—which also results in a lower dry matter content of the vegetation. Heady (1957) noted that the effect of cages was related to temperature and was relatively greater in summer owing to rapid growth. If cages are regularly moved to determine net primary herbage production accurately, it is not likely that changes will occur in floristics or soil conditions. This is important as Peterson *et al.* (1956) have shown a reduction in forage yield of 20% due to soil compaction by grazing animals under certain soil conditions. Floristic changes are equally undesirable.

4.84 The calculation of net primary aerial production within the limits of the errors described owing to changed microenvironment is simple. The weight of herbage produced is the difference between the biomass at the commencement of the period, i.e. outside the cage, and the biomass inside the cage at the end. This figure will of course include no primary production channelled into either invertebrate herbivores or dead plant material unless the dead material has been assessed at each harvest, and Wiegert and Evans (1964) technique applied.

4.85 The calculation of herbage intake by the large animals grazing the area is an important ecosystem parameter of considerable nutritional and economic significance and this opportunity should be taken to calculate it. The above-ground herbage should be cut inside and outside the cage at each single sampling date. Providing the period of protection is low, calculation of intake is as follows:

$$\text{Intake} = \frac{\text{Weight of herbage}}{\text{inside (protected)}} - \frac{\text{Weight of herbage}}{\text{outside (grazed)}}$$

This formula over-estimates forage consumption owing to the increased relative growth rate of the caged area.

If, however, the grazing period is long, it is possible to use a formula suggested by Linehan *et al.* (1952) for use on intensively grazed swards:

$$\text{Intake} = (c - f) \times \frac{(\log d - \log f)}{(\log c - \log f)} \text{ *}$$

where c = Weight of herbage outside cage at previous cut.
$\quad\ f$ = Weight of herbage outside cage this cut.
$\quad d$ = Weight of herbage inside cage this cut.

If an accurate assessment of the net primary production entering the vertebrate herbivores is essential, animal methods are required. These methods always involve a measurement of dung output and this itself is an important parameter which must be measured in an integrated study of the grassland ecosystem. The only practical method for measuring dry matter intake is based on the following equation:

$$\frac{\text{Dry matter intake}}{\text{(g./day)}} = 100 \times \frac{\text{fecal output (g./day)}}{\text{indigestibility of dry matter (\%)}}$$

Two values are therefore required, the indigestibility (digestibility) of the consumed herbage and the dung output. Dung output in free-ranging animals is not easy to measure, but the following methods may be used.

The entire fecal output can be collected in bags carried by the animal. This is a method likely to prove unworkable if *wild* herbivores are the principal large consumer. If total collection is impracticable, it is necessary to use an external indicator fed daily which can be recovered quantitatively in the feces. A great deal of literature is available on this method and interested

---

* This formula is based on the assumption that both herbage growth rate and grazing intensity are proportional to the standing crop at any instant of time.

workers should consult for example the review by Reid, J. T., in "Pasture and range research techniques" (1962).

It is also necessary to calculate the digestibility of the *consumed* herbage which provides many sampling problems. There are two main approaches, neither completely satisfactory owing to the selectivity which many herbivores show when grazing. The classical method involves the use of fecal indicators such as nitrogen (Raymond *et al.,* 1954) (Lancaster, 1949 and 1954) or plant chromogens (Reid *et al.,* 1952) which can be related to digestibility measured directly in a metabolism crate using the animal species concerned and the herbage available. More recently, digestibility in ruminants has been determined in vitro (e.g. Alexander and McGowan, 1961; Dent, 1963; Tilley and Terry, 1963) and if a suitable sample of the herbage grazed can be obtained, i.e. by selective clipping or oesophageal fistulation (Bath *et al.,* 1956; Weir *et al.,* 1959) the method combined with measurement of faecal output is of great value. It is likely that similar techniques can be developed for most animal groups and should be considered if grazing animals are important components of the ecosystem.

Although the methods described in above are not of course primary productivity methods, they illustrate the complexity of the apparently simple grazed grassland ecosystem and the need to combine primary and secondary studies into an integrated whole. They again emphasise the importance of the ecosystem concept discussed in Section 2.25.

5

Non Harvest Methods for Determining Net Primary Production

5.1 Introduction

Net primary production can be measured by other methods than that of harvesting the plant material at intervals although this is the most usual and direct of the methods.

Net primary production represents the carbohydrate material elaborated by the plant by the photosynthetic processes (less that respired). It is, therefore, possible to measure net primary production by measurement of apparent photosynthesis, usually by the measurement of carbon dioxide uptake. There are, however, many problems to the interpretation of this type of physiological measurement to give statements on annual or even weekly productivity and more useful estimates await detailed work of the type described below.

5.2 Use of gaseous exchange techniques

Many methods of measuring net photosynthesis have been described, mainly based on carbon dioxide uptake (Billings *et al.* (1966), Scott and Billings (1964), Hadley and Bliss (1964), Bliss and Hadley (1964)) and several others. These techniques have all used the continually recording technique of infra-red gas analysis. The infra-red gas analyser can be adapted to field measurement by using a generator and is therefore of particular value in grassland studies remote from mains power. All the studies used perspex or plexiglass chambers enclosing single plants or single shoots or on occasions a portion of the entire plant community. With chambers of this type it is possible to use other methods of measuring carbon dioxide concentration. Possibly useful ones are described by Wallis and Wilde (1957).

Monteith (1962), (1963), (1964) has used aerodynamic gas exchange methods which avoid the difficulties of enclosing plants in chambers. These methods which rely on measurements of the carbon dioxide profile down the plant community are of particular value, but their detailed description is beyond the scope of this handbook. They have the disadvantage of requiring a power supply and not being completely reliable if there is any disturbing feature of the environment causing turbulence. Gaseous exchange methods have a particular value in allowing an estimate of gross primary production. This has particular relevance to the description of the characteristics of the ecosystem, the significance of which has been indicated in Section 2.25.

5.3 Chlorophyll techniques

The relationship between the chlorophyll content of natural communities and their net primary production has been broadly demonstrated by Bray (1960) and by Brougham (1960) more specifically for the grassland biome. Although the relationship is not simple and requires further investigation before its use as an index of productivity can be applied universally, it is suggested that the various IBP programmes provide an ideal opportunity for establishing the relationship more exactly and chlorophyll should, therefore, be measured in as many IBP investigations as possible.

5.4 Radioactive tracer techniques

5.41 These techniques have increased greatly in recent years and may be applicable in IBP studies in two ways. Firstly, by the use of $C^{14}O_2$ to study photosynthesis and the transport of carbohydrates within the plant system. This technique can be used to measure net or gross primary production over a short time interval and Peterken and Newbould (1966) have used a technique for woody species similar to that used in grassland by Nasyrov *et al.* (1962). The technique involves a transparent assimilation chamber which fits over $0\cdot1$ m.^{2} of grass which is then filled with a selected $C^{14}O_2$/air mixture. (Weiser *et al.* (1962) have described a simple device for introducing $C^{14}O_2$ to a photosynthetically active plant at a constant predetermined rate.) After allowing the plant to photosynthesise in the $C^{14}O_2$ atmosphere this is harvested, freeze-dried and counted in the normal way to determine the quantity of $C^{14}O_2$ which has been taken up. The count rate is calibrated against net primary production allowing an accurate estimate of net photosynthesis. The method has great value in non-powered sites. If physiological studies of translocation of assimilates are considered, the $C^{14}O_2$ technique offers many possibilities. Workers planning to use this approach with auto-radiography should consult Crafts and Yamaguchi (1964) for details of methodology and the main problems involved.

Examples of the use of radio-isotopes in the study of root systems is given in Section 4.66.

5.42 The other possible use of radiotracer techniques is in a study of mineral cycling often carried out in conjunction with a primary productivity study. This has a particular value in the type of ecosystem study which IBP is most concerned with. Relevant methodology and its use in ecology are described in, for example, Reichle (1967), although the latter techniques were mainly developed for invertebrate studies. A more general discussion of the role of isotopes in ecology is given in Schultz and Klement (1963) and Odum (1959).

5.5   Non-destructive index techniques

Harvest methods although of universal applicability involve the destructive measurement of the increment of the plant community. However, the incremental method can be applied non-destructively by making measurement at time intervals of some characteristic related to dry weight. Such a parameter could be leaf length (i.e. its elongation in a given time) or the increase in diameter of a tussock. (See, for example, Scott (1961) and Mark (1965).) Methods of this type have found considerable application and although inaccuracies are apparent they can provide information which may not be obtainable in any other way. The method although widely used for indicating plant vigour has rarely been used in production studies, Bliss (1966) however, utilised stem and leaf elongation techniques in a study on arctic/alpine species in open situations in conjunction with clipping techniques for denser communities.

5.6   Electronic techniques

Fletcher and Robinson (1956) suggested the use of electronic methods for the non-destructive estimation of herbage weight and Cambell *et al.* (1962) developed this approach. Other workers (for example Alcock (1964), Hyde and Lawrence (1964), and Johns *et al.* (1965)) have further developed the instrument to a stage where it can be used for ecological studies. The principle (basically similar in each case) is that the capacitance of a probe varies with the amount of herbage contained within it in a linear fashion. The probe consisting usually of a number of legs is placed on the grassland sward and the capacitance change recorded as an ammeter reading. The exact factors involved have not been fully elucidated but surface moisture at normal levels affects the reading obtained very little, although there is a close relationship between the reading and total water contained in the herbage. The presence of standing dead material of varying moisture levels and sward density also affects the reading. Although further work is desirable the instrument has been widely used on sown swards (Alcock and Lovatt, 1967) on range vegetation (Van Dyne, 1966) and a more sensitive instrument used on semi-natural mountain grasslands in N. Wales.

An interesting recent development in non-destructive herbage standing crop measurement is the beta-ray attentuation device described by Mott *et al.* (1965). This is in the early developmental stage but offers considerable prospects of a quick, reliable method for herbage measurement of a similar non-destructive type to the electronic capacitance meter.

These non-destructive measures should allow an increase in the samples taken owing to increased speed of recording (although decreasing accuracy). They also give the grassland worker the ability to describe growth curves of an intact grassland community.

The chief disadvantage of the methods is that as yet there is no critical understanding of the factors operating to change the capacitance or the beta-ray attenuation. Although there appears to be a good relationship between capacitance change and several parameters of the herbage yield this lack of knowledge is obviously undesirable and further work is required to more fully understand the mechanism of operation of these very useful techniques.

6

Chemical Analysis of the Plant Material

6.1 Drying of plant material

6.11 Plant material for analysis requires drying, as most chemical analyses are only relevant to material in the dried ground state and all production calculations are based on dry weight or calorific content based on dry weight. In most cases drying in an air oven at temperatures from 80° C.—100° C. will be acceptable, the exact temperature varying with circumstances. The aim must be to prevent the respiratory enzymes functioning in the cut herbage by dehydration as rapidly as possible without causing the loss of volatile organic components. Suitable forced draught ovens have been devised of which the principle features are a downward directed hot air stream ducted to all points of the oven and a system to recirculate air during the initial drying stages. This allows high temperatures to be quickly obtained. (Grassland Research Institute Staff, 1961.) Whatever method is used it should be described as fully as possible in subsequent publications. Several other oven types have been described which are useful in covering a range of sophistication. Hofman (1965) describes the use of a commercial microwave cooker to dry herbage rapidly without raising the sample temperature above 60° C. Greenhill (1960) gives a useful review of methods available and the technique suggested is drying in a vacuum oven at 40° C. over P_2O_5. Oven drying at 80° C. is, however, described as a useful alternative. There is a great necessity for drying facilities suitable for use in field situations. In some cases (e.g. dry tropical climates) air drying to equilibrium moisture content may be acceptable although in every case more standard methods are desirable. Isaacs and Wiant (1959) utilised tractor exhaust gases to dry crop samples very quickly and it is possible that portable methods of this type using the internal combustion engine may be useful for the drying of herbage. In all investigations in which it is necessary to use methods of doubtful repeatability (i.e. air drying) the samples should be stored until drying at a standard temperature can be carried out. In such circumstances it may be better to use published calorific values (e.g. Golley, 1961, Bliss, 1962, Smith, 1967) rather than to determine them on inadequately dried samples.

6.12   If it is essential that chemical changes be kept to a minimum, freeze drying should be employed (Davies *et al.*, 1948, Bath and Budd, 1961). The technique, which is usually very limited in the size of sample it can take, is particularly useful for drying animal or plant material high in fats. In such cases if freeze drying facilities are not available, consideration should be given to the use of a vacuum oven of the type described by, for example, Greenhill (1960).

It may be necessary to preserve fresh samples for either carbohydrate or chlorophyll determinations. The exact methods for this will vary and be dependent on circumstances. In well developed programmes deep freeze units should be available, supplemented by solid CO_2 freezing mixtures for use in the field. Care is necessary to ensure rapid freezing and death, although in the small sample weights envisaged (20 g.—30 g. fresh weight) this should not be a particularly serious problem.

6.2   Grinding of plant material

Grinding is normally carried out in a hammer mill in which the dried sample is subjected to repeated beating until the particles are small enough to pass through a metal screen into a cloth bag or porous container. Such hammer mills can be obtained in a range of sizes and will deal with quantities as low as 5 g. of material. The process of grinding which appears simple, is subject to considerable error and certain basic rules must be observed.

The entire sample should be milled if possible and subsampling carried out on the ground material. If this is not possible, there should be very careful subsampling of the dried cut vegetation prior to grinding.

Great care must be taken to clean out the cloth bag or porous container between samples and to ensure that there are no losses of fine particles either through the bag and its fastening during grinding or subsequently. Such particles are usually high in calorific value and nitrogen content and loss would result in a considerable underestimate of these components in the subsequent chemical analysis. Recent developments utilising completely enclosed chambers with a hardened steel disc to crush the dried herbage (swing mills) have proved satisfactory for grinding small samples. Lieth (in press) discusses grinding in some detail together with sample preparation for bomb calorimetry.

6.3 Calorific value determination

In primary productivity investigations the worker is interested in the calorific values of the various vegetation components, as the development of meaningful energy flow and productivity equations demands theoretical background, best developed in terms of fixed energy content (see for example Skellam, 1967).

The determination of such values is a relatively well known operation involving the use of bomb calorimetry. Several commercial makes of bomb calorimeter are available and the main precautions are documented in several publications (e.g. Grassland Research Institute Staff, 1961). (Raymond *et al.*, 1957). Lieth (in press). It is important to emphasise that ground herbage samples must be pelleted before calorimetry to prevent scattering during burning. The bomb calorimeters mentioned above are for samples of between 1 and 2 gms. of plant material. If considerably less material is available the microbomb calorimeter described by Phillipson (1964) should be used. Maciolek (1962) describes a wet oxidation technique using dichromate which allows the estimation of organic carbon and nitrogen. This technique utilising conversion factors is less accurate than direct methods using bomb calorimeters but is applicable to very small samples indeed (0·07—0·7 mg. organic matter).

6.4 Other analyses

Although calorific value is the most usual and necessary of the analyses performed in productivity studies, it is clear that a considerable increase in academic and practical significance can be obtained if other chemical analyses are performed. It should be possible, therefore, to perform certain basic analyses in grassland ecosystem studies. The most useful analyses are total ash, sodium, potassium, phosphate, calcium and magnesium, together with organic carbon, nitrogen and possibly normal acid (or crude) fibre, soluble carbohydrates and chlorophylls. Methods for these analyses are well documented and publications such as The Association of Official Agricultural Chemists (1965) and Chapman and Pratt (1961) or other appropriate textbooks should be consulted. Deriaz (1961) gives a useful routine method for the main carbohydrate fractions of dried herbage which provides very important additional information particularly if the study includes grazing ruminant herbivores.

Methods for chlorophyll determination have been given by Bray (1960) using constants suggested by Arnon (1949). These methods have the advantage of requiring a single extraction in 90° acetone (i.e. including the water in the plant material). This is followed by spectrophotometric measurement of the extract. This is a considerable saving on the more normal methods which required transferring the pigments to an ether phase before measurement. Freshwater and marine biologists have put a considerable effort into determining chlorophylls and carotenoids. Their techniques and constants should be tested in terrestrial situations and possibly substituted for the Arnon constants. Parsons and Strickland (1963) and Lorenzen (1967) give the appropriate methodology. The P.F. manual has a very full account of chlorophyll techniques and should be consulted.

Total ash can be determined by heating in a muffle furnace at 520° C. for several hours. The residue following bomb calorimetry, however, also provides an estimate of ash content. This measure is particularly important where the dry weight of a plant species increases appreciably over the growing season by accumulation of silica or other inorganic elements in the tissue.

Carbon can be determined by the method of Belcher and Ingram (1950). This is a rapid technique which has the advantage of small sample size.

Another useful parameter if ruminants are important and which can be obtained *in vitro* is digestibility. Tilley and Terry (1963) give details of a method using rumen liquor which has been widely used and developed since its inception. Such developments are described by Alexander and McGowan (1961), Dent (1963) and Rogers and Whitmore (1966) for example which give many useful references.

In all cases the advice of a specialist analyst should be obtained and the methods used recorded in subsequent publication.

7

Results

7.1 Units

The actual recording of data in the field is of course a matter for local decisions but of extreme importance. Care is necessary that field recorders utilise internationally recognisable units and terms to allow maximum exploitation of the data, For this reason the following rules must be observed.

All units should be metric (i.e. *measured in metric units*) although the local equivalents can also be given if considered necessary. Net primary production will normally be expressed as grams/metre^2 in the first instance which can be converted to kgs./ha by multiplication by 10. (This unit is generally used in agricultural experimentation and can be converted to lbs./acre by multiplication by 0·89.)

Energy content should be expressed in calories (gram calorie) or in units of 1000 cals. termed kilo calorie (k. cal.). Net primary production should be quoted in g./m.^{2} or cal./m.^{2}/day or appropriate time unit *carefully defined*. Much confusion has been caused by the practice of expressing an annually measured net primary production on a daily basis without specifying the number of days used in its calculation (i.e. entire year or growing season only). In advanced stations measuring at much closer intervals such confusion should not arise. Total net primary production should be expressed in cals./metre^2/annum and will obviously be the sum of the monthly production and will therefore only include the growing season although expressed over the entire year.

In integrated studies all energy flow measurements should be on the same area basis or at least be easily converted to the same basic area (by introducing factors of 10 to the units). The amount of inorganic components should be expressed as mgs./g. dry matter of the element and not converted to compounds such as P_2O_5 or CaO.

7.2   Errors

In every study the errors associated with the measurements should be clearly stated, preferably by precise statistical methods (derived from standard statistical texts). In some cases, however, such precision may be inappropriate and then the errors should be stated qualitatively with an expression of the investigators' assessment of the accuracies of the methods used. This is an imprecise solution but may permit some comparison with other results. If considerable annual variation in an environmental variable occurs which is likely to affect all aspects of the ecosystem, it is necessary to continue the investigation for a sufficient period to obtain data during as many of the variations as possible. However, in such situations it is appreciated that a mean net primary production figure is unlikely to be meaningful biologically and it may be necessary to quote individual years separately. It is important to remember that an estimate albeit imprecise is often of value at least in initial studies and workers should not be inhibited from studying an ecosystem because of high variability in the data.

7.3   Publications

Research covered by this handbook should be reported in the usual way through scientific journals, although in exceptional cases a separate publication may be appropriate, particularly where an integrated programme requires integrated publication. Copies of all publications should be sent to IBP Central Office (7, Marylebone Road, London, N.W.1, United Kingdom). Bulky original data should be stored in a readily accessible form and made available on request.

8

Summary of Procedures

8.1 Ungrazed grassland

The study area should be delimited and fenced.

8.11 Before beginning the study a trial to determine the optimum size, shape and number of sample plots should be undertaken. (Sections 4.2 and 4.3.)

The appropriate size and number of sample plots should be cut at the beginning of the growing season.

The cut vegetation should be stored in polythene bags, and either deep frozen or dried at 100° within a few hours. A sub-sample should be kept deep frozen if possible.

The same number of quadrats should be cut at appropriate or monthly (4 weekly) intervals throughout the growing season, each successive sample cut being taken in the same general area as the first but no closer than 3 m. in tall herbage (less, if necessary, in short grasslands).

If possible on each harvest date, the quantity of dead material should be determined either directly by separation or by the methods discussed in Section 4.53.

At the same time as harvest of above ground vegetation root cores should be taken, the size and depth to depend on previous investigation.

The dried material harvested, both roots and shoots, should have their calorific value determined using standard bomb calorimetry (Section 6.3). This should include samples of dead and green if separated. At least three separate determinations should be made on each component at each harvest date and calorific value expressed as the mean of these determinations. Total chlorophyll should be measured on fresh or frozen rather than dried material (Sections 6.2, 6.4). A general survey of the small herbivore fauna should be undertaken to determine possible losses of net primary production which will not be determined by the harvest method.

8.12 Calculations. Using the method outlined above net primary herbage production can be calculated as follows:

If B_1 is the above ground biomass measured at the first sampling period (time t_1) and B_2 is the biomass at the second sampling period (time t_2) and B_n is the biomass at the nth sampling period at the end of the growing season (time t_n). Then:

Total annual net primary aerial production is given by

$$(B_2 - B_1) + (B_3 - B_2) + \cdots \cdots \cdots + (B_n - B_{n-1})$$

viz. $\sum_{2}^{n-1} (B_n - B_{n-1})$

and the mean daily net primary production is given by

$$\left(\frac{B_2 - B_1}{t_2 - t_1} \right) \quad \text{or} \quad \left(\frac{B_n - B_{n-1}}{t_n - t_{n-1}} \right)$$

for the appropriate period.

If B is expressed as g./m.^{2} then the final total will be g./m.^{2} To convert to calories/m.^{2} each B must be multiplied by the calorific value/gm. of the harvested sample.

Although the problems of root production are great (see Section 4.6) in certain cases, i.e. annual and ephemeral grasslands where all the roots die, a similar calculation utilising changes in root biomass will give the total net root production. This figure added to herbage production gives *net primary production*. In many unfortunately this will not be possible and root production will need to be measured indirectly and the data added to the herbage production. (These calculations do not take into account plant material dying during the growing period and the techniques discussed in Sections 4.52 and 4.72 should be used, (e.g. Wiegert and Evans (1964).)

8.2 Grazed grasslands

8.21 The method outlined below applies where large grazing animals obviously remove a large proportion of the above ground production. It can also apply where small rodents are the only herbivores if the cages are made with wire mesh which excludes them.

The study area should be delimited and provision made for recording the number of large herbivores present in the area. This need only involve visual methods.

At the beginning of the growing season the number of cages decided by a preliminary trial (see Sections 4.2 and 4.3) and designed to exclude the expected herbivores should be randomly distributed in the study area. The cage should enable a sample plot of appropriate size to be cut within the protected area, avoiding edge effects. A similar size and shaped sample plot should be cut in the immediate vicinity of each of the cages (within 1 m.) and the cut area marked (in many swards the effect of cutting will persist and is a sufficient mark itself).

At monthly intervals the cages are moved to a grazed area within 2 m. of their previous cage but on a previously uncut portion of the community. Paired sample plots are cut (after the initiation of the study) one inside the cage protected from grazing and one on the immediately adjacent grazed area (within 1 m. of the cage).

Grazed swards often have a low standing crop and particular care should be taken to standardise cutting heights and to avoid contamination by soil.

The quantity of dead herbages should be determined as in ungrazed grassland (Section 4.7).

The net primary root production is calculated as in Section 4.6 and added to the aerial production to give net primary production.

Calorific value should be determined as detailed in Section 6.3.

Calculations

If B_1 is the above ground biomass inside the cage at time t_1, B_2 is the biomass inside the cage at time t_2 and W_2 the biomass outside the cage at time t_2 and B_n is the biomass inside the cage at time t_n (end of growing season), then:

Total annual net primary aerial production is given by

$$(B_2 - W_1) + (B_3 - W_2) + \cdots\cdots\cdots + (B_n - W_{n-1})$$

viz. $\displaystyle\sum_{2}^{n-1} (B_n - W_{n-1})$

and the mean daily net primary production is given by

$$\frac{(B_n - B_{n-1})}{t_n - t_{n-1}}$$

for the appropriate period.

The calculation of approximate intake by the excluded herbivores is as follows:

Total annual intake (i.e. removal from plant community) is given by

$$(B_2 - W_2) + (B_3 - W_3) + - - - - - - - - - - (B_n - W_n)$$

and for a particular period the mean daily intake is

$$\frac{(B_n - W_n)}{T_n - t_{n-1}}$$

for the appropriate period.

In the above calculations if B and W are expressed as g./m.^{2} then the final total will be g./m.^{2}. To convert to calories/m.^{2} each B and W must be multiplied by the calorific value/g. of the harvested sample.

If the grazing period is long, the modification by Linehan *et al.* (1952) should be applied (Section 4.84) if possible.

The problem of root production is similar in grazed situations to that where grazing animals are not an important feature and if possible the root production should be added to the net primary aerial production to give net primary production.

Part II

Measurement of the Primary Production of Dwarf Shrub Heaths

C. H. GIMINGHAM AND G. R. MILLER

9

Introduction

Dwarf-shrub heaths are especially characteristic of the Scottish Highlands, where extensive areas are maintained by burning for sheep-farming and grouse-shooting. However, they also occur in other oceanic or sub-oceanic parts of the world and in western Europe are widespread on generally acidic soils of low fertility.

Up until the second World War, heathland received rather little attention from ecologists despite its economic importance. Even today, very little study has been made of the annual dry matter production of these communities and hence no recognised techniques have been developed as for grasslands and woodlands. Nevertheless the general approach to measurements of production in heathland should not differ substantially from that outlined for grasslands in the matter of site selection and description, harvesting techniques for grazed and ungrazed sites, chemical analyses, etc. The appropriate sections of Part I should therefore be consulted. These notes are intended as a guide to possible procedures for tackling problems which are peculiar to dwarf-shrub heaths.

Some dwarf-shrub heaths are managed by regular or intermittent burning and consequently the dominant dwarf-shrubs form extensive, more or less even-aged stands with low sampling variation. Others are relatively undisturbed, leading to heterogeneous uneven-aged stands with high variation.

Some experience of measuring the production of dwarf-shrubs has been gained with even-aged stands in lightly grazed or ungrazed situations and these notes will describe methods in use at present at the Nature Conservancy Unit of Grouse and Moorland Ecology, Banchory, Scotland. Only tentative suggestions can be made for unmanaged, uneven-aged stands.

10

Methods of Measuring Production

10.1 Introduction and general procedure

The general procedure for site selection and sample plot location detailed in Part I (Sections 3 and 4.1) should be followed.

10.11 The age of the dominant dwarf-shrubs should be determined approximately from annual ring counts on a series of stems. In the case of managed heaths, these counts can sometimes be checked against records of burning dates. Ring counts on *Calluna vulgaris* may underestimate the age by about two years (Watt, 1955). When working with even-aged stands, a series of samples should be harvested from each major age-class. However, it may be sufficient to confine sampling to stands in the pioneer, building, mature and degenerate phases of growth. Watt (1955) and Gimingham (1960) have described the characteristics of these phases for *Calluna*.

10.12 Enclosure is unnecessary if there is no grazing but is otherwise essential. With light or intermittent grazing, the stand under investigation may be enclosed with a more or less permanent fence, leaving a border of unused ground within to minimise rain-splash and shelter effects. If birds constitute a grazing factor in these circumstances, this may have to be allowed for by approximation or in extreme cases by using completely enclosed, moveable cages. Moveable cages are essential where there is heavy grazing (see Part I).

10.13 Harvesting can be done with shears. The type with a single or multiple cutting blade is adequate but in general strong secateurs have been found more useful, enabling the vegetation within the quadrat to be cut to the desired level with greater accuracy. The vegetation rooted within each quadrat should be clipped to ground level. This presents few problems where the ground is firm with a clearly defined surface, but where a great depth of loose litter or peat has accumulated, an arbitrary level may have to be chosen. One possibility is to relate cutting to the position of the first secondary root.

10.14 The harvested material should be sorted into species before determination of oven-dry weight. When a large quantity of material has to be handled, it may be necessary to reduce this by sub-sampling. Dry matter percentages can be determined from the sub-samples and applied to the fresh weights measured in the field.

10.2 Measurement of aerial production by the difference method

The difference in the average biomass of two or more sets of samples taken over a measured time interval gives an estimate of the apparent net aerial production (see Part I, Section 2.2). This can be converted to true net aerial production by taking into account the quantity of litter which falls from the plants during the period of measurement (see below).

10.21 The frequency of harvesting depends on the conditions under which the study is being done. In cases where production is measured in an ungrazed stand, twice a year—at the beginning and end of the growing season—might be considered adequate. The difference between the average biomass at these times, to which must be added the weight of litter deposited during the growing season, will give an estimate of net aerial production for the year. When production is measured under grazing conditions, movable cages are essential. These should be harvested and moved to fresh locations at not more than monthly intervals, depending on the rate of growth. It is worth noting that the small growth increment occurring over a short period may be detected only by harvesting a much larger number of samples than is required if sampling is done twice a year.

10.22 In even-aged stands of dwarf-shrubs, the number and size of sample plots required to give a statistically precise estimate of apparent net aerial production per year will be less than in uneven-aged stands. Although the exact number and size of samples required depends primarily on the biomass of the vegetation, the variability within the stand and the growth increment, it is suggested that in most even-aged stands a minimum of 20—25 quadrat samples of $\frac{1}{4}$ m.^{2} might be needed if samples are harvested twice a year.

In very old even-aged stands, where the dwarf-shrubs are rank and tending to become prostrate with a consequent strong development of pattern in the community, or in uneven-aged stands, a larger number of $\frac{1}{2}$ m.^{2} samples may be required. In all cases, however, the tests detailed in Part I, 4.3, should be used.

10.23 Some dwarf-shrubs, e.g. *Calluna vulgaris* produce litter more or less continuously throughout the year while in others, e.g. *Vaccinium* spp., litter production is often seasonal. It is necessary to estimate the amount of litter falling from dwarf-shrubs between successive harvests so that adjustment can be made to the data for apparent net aerial production. A possible method has been suggested by Cormack and Gimingham (1964). A number of sampling areas not to be used for harvesting should have a coarse, hairy fabric such as stockinette stretched out at ground level between the stems of the dwarf-shrubs. Litter is held where it falls on such a fabric and the quantity may be determined at monthly intervals. This method may not be practicable in stands with a high density of woody stems. As an alternative, small trays made from perforated zinc could be positioned at ground level in the stand. The number and size of such sampling units should be determined by preliminary trials (see Part I, Section 4.2).

10.3   Measurement of aerial production by the direct method

10.31 This method is particularly suitable for measuring production in even-aged stands. It depends for its success on the fact that the current season's growth increment can be easily recognised on the long, short or lateral shoots of many dwarf-shrubs. Small sub-samples consisting of 3—5 whole shoots cut at ground level (i.e. "tillers") are taken at random from each harvested quadrat. Either immediately or after a period of deep-freeze storage, these are separated by hand into current year's growth, other green material, flowers and/or fruits, dead foliage and shoots, and woody stem. After drying and weighing, the proportions of these components within the sub-sample are applied to the total dry weight harvested to give a direct estimate of the production of new shoots, foliage and flowers or fruits during the period of measurement. There is no need to make a separate estimate of litter fall since the quantities falling from new growth are negligible so long as harvesting is done before the flowers or leaves begin to drop off. The separated sub-samples can be submitted individually for chemical analysis.

10.32 Experience with even-aged stands of *Calluna* has shown that, since all tillers are of more or less the same age and size, the variability in the proportion of current growth in the sub-samples is consistently low, with a standard deviation of 10—20% of the mean (Miller, unpublished data). In uneven-aged stands it may be difficult to obtain a representative sub-sample because of the wide range in the age and size of tillers harvested. This might be overcome by a preliminary sorting of the tillers of each species into various categories according to size, and sub-sampling at random from these.

10.33 This method does not measure the amount of material laid down in woody stems. Hence a separate estimate of the production of woody material must be made when this is required. This may be done by the difference method, taking the difference between the biomass of wood at the beginning and end of a period of measurement. Allowance might have to be made for small woody stems falling from the plants as litter.

10.34 As with the difference method, the frequency of harvesting depends on the circumstances. If there is no grazing and if an estimate of wood production is not required, one set of samples harvested at the end of the growing season will be adequate to measure the production of new shoots, leaves, flowers or fruits. However, it may be desirable to monitor seasonal changes in the proportions of the various components of the dwarf-shrubs by sampling at three monthly intervals—early spring, mid-summer, autumn and mid-winter. Under grazing conditions, movable cages harvested at not more than monthly intervals should be used. Unlike the difference method, the direct measurement of production should be capable of detecting quite small growth increments without resorting to an unreasonably large number of samples.

The size of quadrat to be used should be the same as that recommended for the difference method. However, only about one half of the number of samples may be required to give reasonably precise results. Thus in most homogeneous even-aged stands 10—12 quadrats of $\frac{1}{4}$ m.² might be adequate, with the same number of $\frac{1}{2}$ m.² quadrats in older stands. The number and size of quadrats necessary, particularly in uneven-aged stands should, however, be found by a preliminary trial.

10.4 Measurement of root production

There are few acceptable methods of determining the root production of perennial plants. However, the biomass of roots may be estimated by sampling with a corer from quadrats after harvesting and washing out over a sieve. In general the methods recommended for sampling roots in grasslands and woodlands may be used in dwarf-shrub heaths.

11

Conclusions

The main difficulty in measuring the net aerial production of dwarf-shrubs from the difference in the average biomass of successive samples is that the growth increment may be small in relation to the total bulk of material harvested. This, together with the probability of high variation in the stand, necessitates the harvesting of a large number of samples. The field-work is laborious and time-consuming. Moreover, it is essential to make a separate estimate of litter fall and this injects a third source of sampling error into the measurement. Perhaps the most serious disadvantage of this method is that the final result does not differentiate between the production of potentially edible shoots, leaves and flowers and inedible wood. It is therefore of little value in studies of secondary production.

On the other hand the direct method, although ignoring wood increments, gives a measure of production based on a single sample which is exact for the sample concerned and does not rely on comparison with other samples. This is particularly desirable when productivity measurements are made over short periods in movable cages. The production of wood may be measured separately by the difference method. However, in studies of herbage utilisation by large herbivores this may be considered unnecessary.

Part III
Measurement of the Primary Production of Arid Zone Plant Communities

R. O. SLATYER

12

Introduction

Introduction

The problems encountered in measuring the primary production of arid zone plant communities are generally similar to those for other grassland or woodland communities, but differ in degree because:

(a) Many communities exhibit a high degree of variability in structure and species distribution, which may be non-random and associated with micro-topographic features.

(b) Growth tends to be episodic, dependent primarily on intermittent rainfall.

(c) Growth rates vary markedly with the stage of maturity of community, and, because of slow growth rates, on maturation may take many years.

The first two of these features necessitate more elaborate sampling procedures than would be required for uniform, actively growing vegetation. The third feature influences the experimental objectives, and the investigator must decide at which maturity stage he wishes to make his assessment of productivity. The main implications of these features are discussed in the following chapter.

13
Some Growth Characteristics of Arid Communities

Most arid communities are either desert woodlands, consisting of scattered shrubs or small trees with a ground flora consisting of grasses and ephemeral herbs; shrub communities consisting almost entirely of shrubs; or grasslands. The species are commonly perennial and may be highly xeromorphic evergreen species of little economic value or seasonally active soft grasses which die back to dormant bases during periods of rainless weather.

All these communities exhibit episodic growth behaviour, depending primarily on soil water recharge from significant falls of rain, but they may also show endogenous seasonal growth characteristics which are temperature and photoperiod determined. Thus some species are winter active and others summer active; while some will flower in certain seasons regardless of soil water supply. The growth characteristics of any two communities may be quite different because of these effects, hence an understanding of the phenology of the major component species of a community is a pre-requisite to an understanding of community growth dynamics.

Growth rate will also vary with the stage of growth of the community. If a situation is envisaged in which a community is grown from seed, growth rate, under uniform environmental conditions (including soil water supply), will show characteristic organic growth form, increasing from low levels when leaf area alone may be limiting photo-synthesis, through a period of most rapid growth which may be determined largely by the genetic characteristics of the species, to a stage of declining growth rates as senescence or specific environmental factors becomes limiting.

In arid communities this pattern is still recognisable, even though periods of growth during favourable environmental conditions are interspersed by periods in which there may be a net loss of dry matter. Ultimately, soil water supply limits further leaf area expansion and, in a perennial evergreen community, a quasi-equilibrium situation is set up in which total standing biomass is more or less constant and the net rate of dry matter production is close to zero. This situation is schematically depicted in Figure 4, assuming a similar rainfall and soil water supply situation prevails each year.

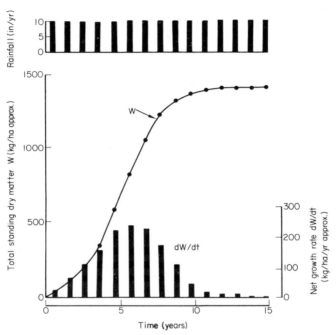

Figure 4. Schematic representation of annual dry matter increments (net growth rate) and total standing dry matter of a perennial evergreen arid community, assuming identical rainfall each year.

Under normal conditions, with fluctuating annual rainfall, the situation is better represented by Figure 5 which shows how total standing biomass and growth rate may vary around the level characteristic of the ecosystem, depending on soil water supply. Thus, during a period of years of above average rainfall, total biomass may increase, while during a period of below average years it may decline. It can be appreciated that, over a period of, say, 5 or 10 years, there may be no significant change in standing biomass, even though there has been a net gain in some years and a net loss in others.

Within years the same pattern of alternate gain and loss is repeated and within seasons it can be most pronounced, instantaneous growth rates some-

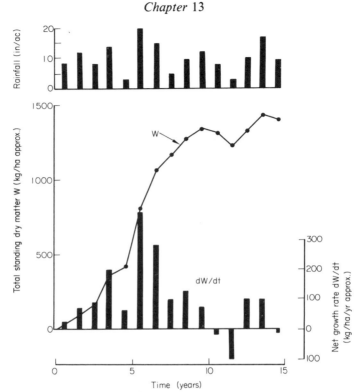

Figure 5. Schematic representation of annual dry matter increments or decrements (net growth rate) and total standing dry matter of a perennial evergreen arid community, assuming fluctuating rainfall totals from year to year.

times being as rapid as those of well watered humid communities, and at other times being strongly negative.

This picture is changed to some degree if material which is shed from the community is also determined, but the overall pattern remains the same, the balance between gross photo-synthesis and gross respiration fluctuating from positive to negative, and lengthy periods of respiratory drain not uncommonly causing an overall loss in total biomass, both standing and shed. However, in most communities there is continued replacement of shed material which, over a period of years, reflects continued, but slow, production.

Within communities which are deciduous, in which there is episodic or seasonal die-back of shoots, or which contain ephemeral annual species, a somewhat different pattern emerges. This generally shows more marked changes in standing biomass between growth and non-growth periods and, in long established communities, continued production at a substantially more rapid rate than in "mature" perennial evergreen communities. This occurs because the dry matter increment which is retained in the community as surface litter, or as standing dead material, following a period of growth, tends to be greater than in the perennial evergreen communities. In turn, this is partly because the communities are effectively at a more "youthful" stage of maturity and respiratory drain from the reduced amount of living material is generally much less. Effectively, therefore, shoot dieback, or leaf-shed, has a rejuvenating effect on the plant community. The reduction in leaf area, and in the volume of living material, reduces both continued soil water extraction and respiration during periods of dry weather and provides a potential for growth and leaf area expansion when soil water is again available.

14

Requirements of the Study Area

Selection of communities for experimental purposes must recognise these patterns and must also recognise that community productivity, on a particular soil in a particular climate, will be influenced by physical and chemical environmental (soil and climatic) factors and by the genetic characteristics of the component species. The objectives of the study must be clearly borne in mind in deciding whether short or long term measurements are to be made, at what stage of maturity the community is to be examined, and whether additional experimental treatments (such as fertiliser application or mechanical defoliation) are to be imposed.

Assuming that the regional or local requirements for the site can be met (i.e. major climatic and soil factors and technical requirements associated with scientific staff and services), the most desirable site features for a study area appear to be those which minimise the problems already outlined. Ideally a site should contain:

(a) An extensive area of vegetation of uniform structure and distribution, preferably containing only one species.

(b) Level topography, so that run off-run on problems are minimised.

(c) Soil of uniform coarse-medium texture, and non-swelling in character.

The layout of the experimental area should resemble that for grassland shown in Part I section. The general rules given there should be followed if at all possible.

The minimum experimental requirements are then:

(a) The stage of maturity of the test community should be categorised and, if appropriate, experimental treatments, such as burning, different degrees of cultivation, reseeding, fertiliser application, etc., should be imposed prior to the experimental period.

(b) Sufficient environmental measurements should be made to enable the major components of the water and energy balance of the ecosystem to be evaluated.

(c) Primary productivity measurements should be made at sufficient frequency to enable short (within season) term changes in productivity to be detected. The frequency of such measurements should not be greater than the length of each growth period.

References

Reference Works and Handbooks

ASSOCIATION OF OFFICIAL AGRICULTURAL CHEMISTS (1965). *Official Methods of Analysis of the Association of Official Agricultural Chemists.* 10th ed. Assoc. Official Agr. Chemists, Washington D.C.

BROWN D. (1954). *Methods of Surveying and Measuring Vegetation.* Commonwealth Bureau of Pastures and Field Crops, Bul. 42, Farnham Royal, Bucks, England.

CAIN S.A. & DE O. CASTRO G.M. (1959). *Manual of Vegetation Analysis.* Harper, New York.

CHAPMAN H.D. & PRATT P.F. (1961). *Methods of Analysis for Soils, Plants and Waters.* University of California.

CURTIS (1959. *The Vegetation of Wisconsin: an Ordination of Plant Communities.* Madison, Wisconsin.

GRASSLAND RESEARCH INSTITUTE STAFF. *Research Techniques in Use at the Grassland Research Institute, Hurley.* Bulletin 45. Commonwealth Bureau of Pastures and Field Crops, Hurley, Berks, England.

GREIG-SMITH P. (1964). *Quantitative Plant Ecology (second edition).* Butterworths, London.

JOINT COMMITTEE (AMERICAN SOC. OF AGRONOMY; AMER. DAIRY SCI. ASSOC.; AMER. SOC. RANGE MAGMT) (1962). *Pasture and Range Research Techniques.* Comstock (Cornell Univ. Press), New York.

JOINT COMMITTEE AMER. SOC. RANGE MGMT & AGRIC. BOARD (1962). *Basic Problems and Techniques in Range Research.* Publication No. 890 Natl. Acad. of Sciences—Natl. Research Council, Washington D.C.

KERSHAW K.A. (1964). *Quantitative and Dynamic Ecology.* Arnold, London.

MORRIS M.J. (1967). *An Abstract Bibliography of Statistical Methods in Grassland Research.* U.S. Forest Service, U.S. Dept. of Agriculture Misc. Publ. 1030.

ODUM E.P. (1959). *Fundamentals of Ecology (second edition).* Saunders, Philadelphia and London.

PETRUSEWICZ K. (ed.) (1967). *Secondary Productivity of Terrestrial Ecosystems (Principles and Methods).* Institute of Ecology, Polish Academy of Science, Warsaw (IBP).

SOUTHWOOD T.R.E. (1966). *Ecological Methods.* Methuen, London.

SCHULTZ V. & KLEMENT A.W. (Jr.) (1963). *Radioecology.* Reinhold Corp., N. York.

Papers

ALCOCK M.B. (1964). *An Improved Electronic Instrument for Estimation of Pasture Yield.* Nature, Lond. 203 : 4951 : 1309.

ALCOCK M.B. & LOVETT J.V. (1967). *The Electronic Measurement of the Yield of Growing Pasture. (I) A Statistical Assessment.* J. Agric. Sc. 68 : 27.

ALDER F.E. & RICHARDS J.A. (1962). *A Note on the Use of the Power Driven Sheep Shearing Head for Measuring Herbage Yields.* J. Brit. Grassl. Soc. 17 : 2 : 101.

ALEXANDER R.H. & McGOWAN M. (1961). *A Filtration Procedure for the in vitro Determination of Digestibility of Herbage.* J. Br. Grassl. Soc., Vol. 16, pp. 275–7.

ARNON D.I. (1949). *Copper Enzymes in Isolated Chloroplasts.* (1) *Polyphenoloxidase in* B. vulgaris. Plant Physiol. 24 : 1.

BATH D.L., WEIR W.C. & TORELL D.T. (1956). *The Use of Oesophageal Fistula for the Determination of Consumption and Digestibility of Pasture Forage by Sheep.* J. Animal Sci. 15 : 1166.

BATH I.H. & BUDD R.T. (1961). *A Laboratory Apparatus for Freeze Drying Herbage Samples.* J. Brit. Grassl. Soc. 16 : 4 : 278.

BECKER C.F. (1959). *Equipment for Harvesting Short-Grass Rangeland Plots.* Agron. J. 51 : 7 : 430.

BELCHER R. & INGRAM S. (1950). *A Rapid Micro Combustion Method for the Determination of Carbon and Hydrogen.* Anal. Chim. Acta. 4. 118.

BILLINGS W.D., CLEBSCH E.E.C. & MOONEY H.A. (1966). *Photosynthesis and Respiration Rates of Rocky Mountain Alpine Plants under Field Conditions.* Amer. Midl. Nat. 75 : 1 : 34.

BLISS L.C. (1962). *Caloric and Lipid Content in Alpine Tundra Plants.* Ecology. 43 : 753–757.

BLISS L.C. (1966). *Plant Productivity in Alpine Microenvironments on Mt. Washington, New Hampshire.* Ecological Monographs 36 : 125.

BLISS L.C. & HADLEY E.B. (1964). *Photosynthesis and Respiration of Alpine Lichens.* Amer. J. Bot. 51 : 8 : 870.

BOYD D.A. (1949). *Experiments with Leys and Permanent Grass.* J. Brit. Grassl. Soc. 4, p. 1–10.

BRAY J.R. (1960). *The Chlorophyll Content of some Native and Managed Plant Communities in Central Minnesota.* Canad. J. Bot. 38 : 3 : 313.

BRAY J.R. (1963). *Root Production and the Esitmation of Net Productivity.* Canad. J. Bot. 41 : 1 : 65.

BROUGHAM R.W. (1960). *The Relationship between the Critical Leaf Area, Total Chlorophyll Content and Maximum Growth Rate of some Pasture and Crop Plants.* Ann. Bot. Lond. New Series 24 : 463.

BROUGHAM R.W. (1962). *The Leaf Growth of* Trifolium repens *as Influenced by Seasonal Changes in the Light Environment.* J. Ecol. 50 : 449.

CAMPBELL A.G., PHILLIPS D.S.M. & O'REILLY E.D. (1962). *An Electronic Instrument for Pasture Yield Estimation.* J. Brit. Grassl. Soc. 17 : 2 : 89.

CAMPBELL R.S. & CASSADY J.T. (1949). *Determining Forage Weight on Southern Forest Ranges.* J. Range Management 2 : 30.

CARLSON C.W. (1965). *Problems and Techniques in Studying Plant Root Systems.* Proc. 9th Int. Grassl. Cong., p. 5.

COLEMAN S.H. (1959). *A Useful Device for Laying Out Forage Production Plots.* J. Range Mgmt. 12 : 3 : 138.

CORMACK E. & GIMINGHAM C.H. (1964). *Litter Production by* Calluna vulgaris (*L.*) *Hull.* J. Ecol. 52, 285–97.

COUPLAND R.T. & JOHNSON R.E. (1965). *Rooting Characteristics of Native Grassland Species in Saskatchewan.* J. Ecol. 53 : 475.

COWLISHAW S.J. (1955). *The Effect of Sampling Cages on the Yields of Herbage.* J. Brit. Grassl. Soc. 6 : 179.

CRAFTS A.S. & YAMAGUCHI S. (1964). *The Autoradiography of Plant Material.* Calif. Agric. Exp. Sta.

DAHLMAN R.C. & KUCERA C.L. (1965). *Root Productivity and Turnover in Native Prairie.* Ecology. 46 : 1–2 : 84.

DAVIES A.W. *et al.* (1948). *A New Technique for the Preparation and Preservation of Herbage Samples.* J. Brit. Grassl. Soc. 3 : 153.

DEFFONTAINES J.P. (1964). *A Method for Appraising the Root System under Grassland.* (*Fr. Eng. summ.*) Fourrages 19 : 91.

DEKLIT C.T. & TALSMA T. (1952). *The Determination of the Activity of Roots by the Use of Radioactive Isotopes.* Landbouwk Tijdscher 64 : 398.

DENT J.W. (1963). *Applications of the Two-Stage in Vitro Digestibility Method of Variety Testing.* J. Br. Grassland Soc., Vo.. 18, pp. 181–9.

DERIAZ R.E. (1961). *The Routine Analysis of Carbohydrates and Lignin in Herbage.* J. Sci. Food Agric. 12 : 152.

DONOVAN L.S. *et al.* (1958). *A Photoelectric Device for Measurement of Leaf Areas.* Canad. J. Pl. Sci. 38 : 4 : 490.

EDWARDS P.J. (1965). *Effect of Storage in Polythene Bags on the Green Weight and Dry Matter Content of Herbage Samples.* S. Afr. J. Agric. Sci. 8 : 2 : 337.

FLETCHER J.E. & ROBINSON M.E. (1956). *A Capacitance Meter for Estimating Forage Weight.* J. Range Mgmt. 9 : 96.

FRIBOURG H.A. (1953). *A Rapid Method for Washing Roots.* Agron. J. 45 : 334.

GIMINGHAM C.H. (1960). *Biological Flora of the British Isles.* Calluna vulgaris (*L.*) Hull. J. Ecol. 48, 455–83.

GREEN J.O. (1949). *Herbage Sampling Errors and Grazing Trials.* J. Brit. Grassl. Soc., 4, 11–16.

GREENHILL W.L. (1960). *Determination of the Dry Weight of Herbage by Drying Methods.* J. Brit. Grassl. Soc. 15 : 1 : 48.

GOLLEY F.B. (1961). *Energy Value of Ecological Materials.* Ecology 42 : 3 : 581.

HADLEY B. & BLISS L.C. (1964). *Energy Relationship of Alpine Plants on Mount Washington, New Hampshire.* Ecol. Monogr. 34 : 331.

HALL N.S., CHANDLER W.F., VAN BAVEL G.H.M., REID P.H. & ANDERSON J.H. (1953). *A Tracer Technique to Measure Growth and Activity of Plant Root Systems.* N.C. Agr. Expt. Stn. Tech. Bull. 101.

HEADY H.F. (1957). *Effect of Cages on Yield and Composition in the California Annual Type.* J. Rnge. Mgmt. 10 : 4 : 175.

HEADY H.F. & VAN DYNE G.M. (1965). *Prediction of Weight Composition from Point Samples of Clipped Herbage.* J. Rnge. Mgmt. 18 : 144.

HILPOLTSTEINER L. (1960). *An Improved Sampling Technique for Grassland Experiments.* Landw. Forsch. 13 : 1 : 42.

HOFMAN M.A.J. (1965). *Microwave Heating as an Energy Source for the Pre-Drying of Herbage Samples.* Pl. Soil 23 : 1 : 145

HUNTER R.F. & GRANT S.A. (1961). *The Estimation of "Green Dry Matter" in a Sample by Methanol-Soluble Pigments.* J. Brit. Grassl. Soc. 16 : 1 : 43.

HYDE F.J. & LAWRENCE J.T. (1964). *Electronic Assessment of Pasture Growth.* Electron. Engng. 36 : 777.

IL'IN S.S. (1961). *Methods of Studying the Root Systems of Plants.* Bot. Z S.S.S.R. 46 : 10 : 1533.

ISAACS G.W. & WIANT D.E. (1959). *A Rapid Drying Oven for Determining the Moisture Content of Crop Samples in the Field.* Quart. Bull. Mich. Agric. Exp. Sta. 41 : 3 : 600.

JACQUES W.H. & SCHWASS R.H. (1956). *Root Development in Some Common New Zealand Pasture Plants.* N.Z. J. Sci. Tech. 37 : 569.

JAGTENBERG W.D. & DE BOER T.A. (1957). *De Bruikbaarheid van Graskooien voor Opbrengstbepalingen.* Landb-Voorl 14 : 12 : 622.

JENKINS H.V. (1959). *An Airflow Planimeter for Measuring the Area of Detached Leaves.* Pl. Physiol. 34 : 5 : 532.

JOHNS G.G., NICOL G.R. & WATKIN B.R. (1965). *A Modified Capacitance Probe for Estimating Pasture Yield.* J. Brit. Grassl. Soc. 20 : 4 : 212.

JONES R.I. (1961). *A Simple Apparatus for Estimation of the Area of Detached Leaves.* S. Afr. J. Agric. Sci. 4 : 4 : 531.

KEMP C.D. (1960). *Methods of Estimating the Leaf Area of Grasses from Linear Measurements.* Ann. Bot. Lond. 24 : 96 : 491.

LANCASTER R.J. (1949). *The Measurement of Feed Intake by Grazing Cattle and Sheep. (I) A Method of Calculating the Digestibility of Pasture Based on the Nitrogen Content of Faeces Derived from Pasture.* New Zealand J. Sci. Tech. 31 : 3.

LANCASTER R.J. (1954). *Measurement of Feed Intake of Grazing Cattle and Sheep. (V) Estimation of the Feed to Faeces Ratio from the Nitrogen Content of the Faeces of Pasture Fed Cattle.* New Zealand J. Sci. Tech. 36 : 15.

LAVIN F. (1961). *A Glass-Faced box for Field Observations on Roots.* Agron. Journ. 53 : 4 : 265.

LIETH H. (in press). *The Measurement of Calorific Values of Biological Material and the Determination of Ecological Efficiency.* UNESCO, Copenhagen. Symposium.

LORENZEN C.J. (1967). *Determination of Chlorophyll and Pheo-Pigments; Spectrophotometric Equations.* Limnol. Oceanogr. 12 : 343.

LINEHAN P.A., LOWE J. & STEWART R.H. (1952). *The Output of Pasture and its Measurement.* J. Brit. Grassl. Soc. 7 : 73.

MACIOLEK J.A. (1962). *Limnological Organic Analyses by Quantitative Dichromate Oxidation.* United States Department of the Interior. Fish and Wildlife Service Research Report 60.

MARK A.F. (1965). *The Environment and Growth Rate of Narrow Leaved Snow Tussock in Otago.* N.Z. Journ. Bot. 3 : 2.

MATCHES A.G. (1963). *A Cordless Hedge Trimmer for Herbage Sampling.* Agron. J. 55 : 3 : 309.

MCGINNIES W.J. (1959). *A Rotary Lawn Mower for Sampling Range Herbage.* J. Rnge. Mgmt. 12 : 4 : 203.

MCKELL C.M., WILSON A.M. & JONES M.B. (1961). *A Flotation Method for Easy Separation of Roots from Soil Samples.* Agron. J. 53 : 1 : 56.

MONTEITH J.L. (1962). *Measurement and Interpretation of CO_2 Fluxes in the Field.* Neth. J. Agric. Sci. 10 : 5 : 334.

MONTEITH J.L. (1963). *Gas Exchange in Plant Communities*. In· Environmental Control of Plant Growth, ed. L.T. Evans, New York.

MONTEITH J.L., SZEICZ G. & YABUKI K. (1964). *Photosynthesis and Carbon Dioxide Flux within Crops*. J. Appl. Ecol. 1 : 321.

MOTT G.O., BARNES R.F. & RHYKERD C.L. (1965). *Estimating Pasture Yield in situ by Beta-Ray Attenuation Techniques*. Agron. J. 1965 : 57 : 5 : 512.

MUZIK T.J. & WHITWORTH J.W. (1962). *A Technique for the Periodic Observation of Root Systems in situ*. Agron. J. 1962. 54 : 1 : 56.

NASYROV JU.S., GILLER JU.E., LOGINOV M.A. & LEBEDEV V.N. (1962). *The Use of C^{14} for Studying the Photosynthetic Balance of Plant Communities*. Bot. Z S.S.S.R. 47 : 1 : 96.

NEILSON J.A. (1964). *Autoradiography for Studying Individual Root Systems in Mixed Herbaceous Stands*. Ecology. 45 : 3 : 644.

NEWBOULD P. (1967). *Methods for Estimating the Primary Production of Forests*. IBP Handbook No. 2, Blackwell, Oxford.

ORCHARD B. (1958). *Measurement of Leaf Area*. Report. Rothamsted Exp. Sta., Harpenden, 1959.

ORCHARD B. (1961). *Measurement of Leaf Area*. J. Exp. Bot. 1961 12 : 36 : 458.

PARSONS T.R. & STRICKLAND J.D.H. (1963). *Discussion of Spectrophotometric Determination of Marine Plant Pigments, with Revised Equations for Ascertaining Chlorophylls and Carotenoids*. J. Mar. Res. 21 : 155.

PECHANEC J.F. & STEWART G.· (1940). *Sagebrush-Grass Range Sampling Studies*. Agron. J. 32 : 669.

PETERKEN G.F. & NEWBOULD P.J. (1966). *Dry Matter Production by Ilex Aquifolium (L) in the New Forest*. J. Ecol. 54 : 1 : 143.

PETERSEN R.G., WOODHOUSE W.W. & LUCAS H.L. (1956). *The Distribution of Excreta by Freely Grazing Cattle and its Effect on Pasture Fertility*. Agron. Journ. 48 : 444.

PHILIP J.R. (1965). *The Distribution of Foliage Density with Foliage Angle Estimated from Inclined Point Quadrat Observations*. Aust. J. Bot. 13 : 2 : 357.

PHILLIPSON J. (1964). *A Miniature Bomb Calorimeter for Small Biological Samples*. Oikos 15 : 1 : 130.

PILET P.E. & MEYLAN A. (1958). *A Method of Measuring Leaf Area*. Ber. Schweiz. bot. Ges. 68 : 307.

RAYMOND W.F., KEMP C.D., KEMP A.W. & HARRIS C.E. (1954). *Studies in the Digestibility of Herbage. (IV) The Use of Faecal Collection and Chemical Analysis in Pasture Studies. (b) Faecal IndexMethods*. J. Brit. Grassl. Soc. 9 : 69.

RAYMOND W.F. et al. (1957). *An Automatic Adiabatic Bomb Calorimeter*. J. Sci. Inst. 34 : 501.

REICHLE D. (1967). *Radiosotope Turnover and Energy Flow in Terrestrial Isopod Populations*. Ecology. 48 : 3 : 351.

REID J.T., WOODFOLK P.G., HARDISON W.A., MARTIN C.M., BRUNDAGE A.L. & KAUFMAN R.W. (1952). *A Procedure for Measuring the Digestibility of Pasture Forage under Grazing Conditions*. J. Nutr. 46 : 255.

REPPERT J.N., REED M.J. & ZUSMAN P. (1962). *An Allocation Plan for Range Unit Sampling*. J. Rnge. Mgmt. 15 : 4 : 190.

RODER W. (1959). *Method and Results of Testing Root Production of Genetically Valuable Single-Plant Progeny of Sheeps' Fescue when Sown in Rows.* Z. f. landw. Vers.-u. Untersuch Wes. 5 : 2 : 122.

ROGERS H. H. & WHITMORE E.T. (1966). A *Modified Method for the in vitro Determination of Herbage Digestibility in Plant-Breeding Studies.* J. Br. Grassland Soc., 21, 2.

SKELLAM J.G. (1967). *In: Secondary Productivity of Terrestrial Ecosystems (Principles and Methods).* Institute of Ecology, Polish Academy of Science, Warsaw.

SCOTT D. (1961). *Methods of Measuring Growth in Short Tussocks.* N.Z. J. Agric. Res. 4 : 3–4 : 282.

SCOTT D. & BILLINGS W.D. (1964). *Effects of Environmental Factors on Standing Crop and Productivity of an Alpine Tundra.* Ecol. Monog. 34 : 3 : 243.

SHIMADA Y. (1959). *Statistical Studies on the Design of Yield Survey and Field Experiment in Natural Grassland.* (3) *Estimation of Yield Especially with Reference to Size and Shape and Replication of Field Experimental Plot in Natural Zoysia Grassland.* Sci. Reps. Res. Inst. Tohuku Univ., D. 1959. 10 No. 2 : 87.

SMITH A.D. & SHEETS A.M. (1960). *A Technique for Studying Forage Removal by Game and Livestock.* J. Range Mgmt. 13 : 3 : 151.

SMITH D.R. (1967). *Gross Energy Value of Above Ground Parts of Alphine Plants.* J. Range Mgmt. 20 : 3 : 179.

TANSLEY A.G. (1935). *The Use and Abuse of Vegetational Concepts and Terms.* Ecol. 16 : 284.

TATARINOVA N.K. (1961). *The Longevity of the Roots of Meadow Grasses.* Bot. Z S.S.S.R. 46 : 7 : 925.

THILENIUS J.F. (1966). *An Improved Vegetation Sampling Quadrat.* J. Rnge. Mgmt. 19 : 1 : 40.

TILLEY J.M.A. & TERRY R.A. (1963). *A Two-Stage Technique for the in vitro Digestion of Forage Crops.* J. Br. Grassland Soc., 18, 104–11.

TROUGHTON A. (1957). *The Underground Organs of Herbage Grasses.* Bulletin 44. Commonwealth Agricultural Bureau.

TROUGHTON A. (1958). *Studies on the Roots and Storage Organs of Herbage Plants.* Welsh Pl. Breeding Sta. Rep., 1950–56. Aberystwyth. (Also Herb. Abs. 22 : 981.)

TROUGHTON A. (1959). *Studies on the Roots and Storage Organs of Herbage Plants.* Rep. Welsh Pl. Br. Stn. Abs., 1958. 60.

TROUGHTON A. (1960). *Further Studies on the Relationship Between Shoot and Root Systems of Grasses.* J. Brit. Grassl. Soc. 15 : 1 : 41.

TROUGHTON A. (1961). *Root Studies.* Rep. Welsh Pl. Breed. St., 1960. 61–3.

UPCHURCH R.P. (1951). *The Use of Trench Wash and Soil-Elution Methods for Studying Alfalfa Roots.* Agron. J. 43 : 552.

VAN DYNE G.M. (1960). *A Method for Random Location of Sample Units in Range Investigations.* J. Range Mgmt. 13 : 3 : 152.

VAN DYNE G.M. (1966). *Use of a Vacuum-Clipper for Harvesting Herbage.* Ecology. 47 : 4 : 624.

VAN DYNE G.M., VOGEL W.G. & FISSER H.G. (1963). *Infuence of Small Plot Size and Shape on Range Herbage Production Estimates.* Ecology. 44 : 4 : 746.

WAGNER R.E., HEIN M.A., SHEPHERD J.B. & ELY R.E. (1950). *A Comparison of Cage and Mower Strip Methods with Grazing Results in Determining Production of Dairy Pastures.* Agron. J. 42 : 487.

WALLIS G.W. & WILDE S.A. (1957). *Rapid Method for the Determination of Carbon Dioxide Evolved from Forest Soils.* Ecology. 38 : 359.

WATT A.S. (1955). *Bracken versus Heather, a Study in Plant Sociology.* J. Ecol. 43, 490–506.

WEAVER J.E. (1958). (a) *Summary and Interpretation of Underground Development in Natural Grassland Communities.* Ecol. Monogr. 28 : 1 : 55.

WEAVER J.E. (1958). (b) *Classification of Root Systems of Forbs of Grassland and a Consideration of Their Significance.* Ecology. 39 : 3 : 393.

WEIR W.C., MEYER J.H. & LOFGREEN G.P. (1959). *The Use of the Oesophageal Fistula, Lignin and Chromogen Techniques for Studying Selective Grazing and Digestibility of Range in Pasture by Sheep and Cattle.* Agron. J. 51 : 235.

WEISER C.J., WANG C.H. & BLANEY L.T. (1962). *A Simple Apparatus for Feeding $C^{14}O_2$ to Photosynthetically Active Plants at a Constant Predetermined Rate.* Proc. Amer. Soc. Hort. Sci. 80 : 661.

WESTFALL S.E. & DUNCAN, D.A. (1961). *The San Joaquin Cage.* J. Range Mgmt. 14 : 6 : 335.

WIEGERT R.G. (1962). *The Selection of an Optimum Quadrat Size for Sampling the Standing Crop of Grasses and Forbs.* Ecology. 43 : 1 : 125.

WIEGERT R.G. & EVANS F.C. (1964). *Primary Production and the Disappearance of Dead Vegetation on an old field in S.E. Michigan.* Ecology. 45 : 1 : 49.

WIEGERT R.G. (1967). *In Secondary Productivity of Terrestrial Ecosystems (Principles and Methods).* Institute of Ecology, Polish Academy of Science, Warsaw.

WILLARD C.J. & McCLURE G.M. (1932). *The Quantitative Development of Tops and Roots in Blue Grass with an Improved Method of Obtaining Root Yields.* Agron. J. 24 : 509.

WILLIAMS T.E. & BAKER H.K. (1957). *Studies in the Root Development of Herbage Plants.* (1) *Techniques of Herbage Root Investigations.* J. Brit. Grassl. Soc. 12 : 49.

WILSON J.W. (1960). *Inclined Point Quadrats.* New Phytol. 59 : 1–8.

WILSON J.W. (1963). *Estimation of Foliage Denseness and Foliage Angle by Inclined Point Quadrats.* Aust. J. Bot. 11 : 1 : 95.

WILSON J.W. (1965). *Point Quadrat Analysis of Foliage Distribution for Plants Growing Singly or in Rows.* Aust. J. Bot. 13 : 3 : 405.

WILSON J.W. (1965). *Stand Structure and Light Penetration. (1) Analysis by Point Quadrat.* J. Appl. Ecol. 2 : 2 : 383.

Recent and additional references

COCHRAN W.G. & COX G.M. (1957). *Experimental Designs.* 2nd Ed. Wiley, New York.

FEDERER W.T. (1955). *Experimental Design.* Macmillan, New York.

VAN DYNE G.M. (1969) (Ed.). *The Ecosystem Concept in Natural Resource Management.* Academic Press, New York.

Further Reading

ALEKSEENKO L.N. (1959). *Method for Determining Leaf Area of Herbage Plants.* Doklady Vsesojuz. Akad. S-H. Nauk. 24 : 9 : 27.

ALEXANDER C.W., SULLIVAN J.T. & McCLOUD D.E. (1962). *A Method for Estimating Forage Yields.* Agron. Journ. 54 : 5 : 468–469.

ARMY A.C. & SCHMID A.R. (1942). *A Study of the Inclined Point Quadrat Method of Botanical Analysis.* Amer. Soc. Agron. Journ. 34 : 3 : 238–247.

BAKER D.N. & MUSGRAVE R.B. *Photosynthesis under Field Conditions. (V) Further Plant Chamber Studies of the Effects of Light on Corn (Z. mays L.).* Crop Sci. 4 : 127 : 131.

BAKHUIS J.A. (1960). *Estimating Pasture Production by the Use of Grass Length and Sward Density.* Neth. Jour. Agr. Sci. 8 : 3 : 211–224.

BOX T.W. (1960). *Herbage Production in Four Range Plant Communities in S. Texas.* J. Rnge. Mgmt. 13 : 2 : 72.

BRANSON F.A. & PAYNE G.F. (1958). *Effects of Sheep and Gophers on Meadows of the Bridger Mountains of Montana.* J. Range Mgmt. 11 : 4 : 165.

ELLISON L. & HOUSTON W.R. (1958). *Production of Herbaceous Vegetation in Openings and under Canopies of Western Aspen.* Ecology 39 : 2 : 337.

FILIPPOVA L.A. (1959). *Daily and Seasonal Changes in the Intensity of Photosynthesis in Plants of the Eastern Pamir.* Trudy Bot. Inst. V.L. Kamarova. Series 4 : Exp. Bot., No. 13. 64–90 (English summary).

FRENCH M.H. (1961). *Problems associated with Cutting and Weighing Techniques for Measuring Tropical Herbage Productivities.* Turrialba 11 : 1 : 4.

GOLLEY F.B. (1965). *Structure and Function of an Old Field Broom Sedge Community.* Ecol. Monog. 35 : 1 : 113.

GOLUBEV V.N. (1963). *Method for Determining Absolute Productivity of the Above Ground Part of the Herbage Cover of Meadow Steppe.* Bot. Z S.S.S.R. 48 : 9 : 1338.

GRUNOW J.O. & GINKEL B.VAN (1965). *Determination of the Bulk Utilization of Different Veld Grasses under Grazing Conditions by Means of an Individual Tuft Clipping Method.* S. Afr. J. Agric. 8 : 2 : 487.

HECHT H. (1960). *The Estimation of the Area of Single Leaves (Germ.).* Bayer. ladnw. Jahrb. 37 : 4 : 479.

HURD R.M. & POND F.W. (1958). *Relative Preference and Productivity of Species on Summer Cattle Ranges, Big Horn Mountains, Wyoming.* J. Rnge. Mgmt. 11 : 3 : 109.

KOELLING M.R. & KUCERA C.L. (1965). *Productivity and Turnover Relationships in Native Tallgrass Prairie.* Iowa St. J. Sci. 39 : 4 : 387.

KUSUMOTO T. (1958). *Physiological and Ecological Studies on the Plant Production in Plant Communities. (6) On the Plant Production of* Miscanthus *Community in Kagoshima Prefecture.* Memoir Fac. Edu. Kagoshima Univ. 10 (1) : 27.

KUSUMOTO T. (1961). *Studies on the Grass Production in a Grassland in Osumi Peninsula Southern Kyushu.* Misc. Rep. Res. Inst. Natur. Resaur. 54–55 : 101.

LIETH H. (ed.) (1962). *Die Stoffproduktion der Pflanzendecke.* Gustav Fischer Verlag Stuttgart.

LIETH H. (1965). *Ecological Problems in the Investigation of Biological Productivity.* (1) *Introduction, Definitions and Growth Analyses.* Qualitas. Pl. Mater. Veg. 12 : 3 : 241.

McCULLOCH C.Y. (1959). *Effects of Rodents and Rabbits on Estimates of Forage Disappearance.* Proc. Okla. Acad. Sci., 1958. 39 : 202.

NUMATA M. & SHIMADA Y. (1966). *Bibliography of Grassland Ecology in Japan.* Jap. IBP-PT(G), CT.

PASE C.P. (1958). *Herbage Production and Composition Under Immature Ponderosa Pine Stands in the Black Hills.* J. Range Mgmt. 11 : 5 : 238.

PAULSEN H.A. (1960). *Plant Cover and Forage Use of Alpine Sheep Ranges in the Central Rocky Mountains.* Iowa State J. Sci. 34 : 4 : 731.

PAVLYCHENKO T.K. (1937). *Quantitative Study of the Entire Root Systems of Weed and Crop Plants under Field Conditions.* Ecol. 18 : 62.

PEARSALL W.H. & NEWBOULD P.J. (1957). *Production Ecology.* (4) *Standing Crops of Natural Vegetation in the Sub-Arctic.* J. Ecol. 1957. 45 : 2 : 593.

PEARSON L.C. (1965). *Primary Production in Grazed and Ungrazed Desert Communities of Eastern Idaho.* Ecology. 46 : 3 : 278.

RAMAM S.S. (1960). *Root Ecology of Dichanthium Annulatum.* J. Indian bot. Soc. 39 : 12 : 210.

RATCLIFF R.D. & HEADY H.F. (1962). *Seasonal Changes in Herbage Weight in an Annual Grass Community.* J. Rnge. Mgmt. 15 : 3 : 146.

RICKARD W.H. (1962). *Harvest Yields in Natural Plant Communities.* Publ. HW.72500 Hanford Atomic Prod. Op., Washington 159.

ROMNEY D.H. (1960). *Productivity of Pasture in British Honduras.* (*I*) *Natural Pasture.* Trop. Agric. 37 : 2 : 135.

SHISHIDO M. (1960). *Studies on the Range at Kushima District.* Bull. Fac. Agric. Mizazaki 5 : 2 : 93.

SOCAVA V.B. & LIPATOVA W. (1962). *An Experimental Study of the Optimum Productivity of the Above Ground Parts of a Herbage Cover.* Gorskova A. A. Bot. Z S.S.S.R. 47 : 4 : 473. (Russ. Eng. Summ.)

TAMM E. & KRZYSCH G. (1959). *Observations on the Growth Factor CO_2 in the Phytosphere.* Z. Acker-u Pfl. Bau. 107 : 3 : 275.

VAN DYNE G.M. (1960). *A Procedure for Rapid Collection, Processing and Analysis of Line Intercept Data.* J. Rnge. Mgmt. 13 : 5 : 247.

WAGNER R.E., HEIN M.A., SHEPHERD J.B. & ELY R.E. (1950). *A Comparison of Cage and Mower Strip Methods with Grazing Results in Determining Production of Dairy Pastures.* Agron. J. 42 : 487.

WEAVER J.E. & ZINK E. (1946). *Annual Increase of Underground Materials in Range Grasses.* Ecology 27 : 115–27.

WESTLAKE D.F. (1963). *Comparisons of Plant Productivity.* Biol. Rev. 38, 385–425.

WIEGERT R.G. & LINDEBORG R.G. (1964). *A "Stem Well" Method of Introducing Radioisotopes into Plants to Study Food Chains.* Ecol. 45 : 2 : 40(.

WILLARD C.J. & McCLURE G.M. (1932). *The Quantitative Development of Tap Roots in Blue Grass with an Improved Method of Obtaining Root Yields.* Agron. J. 24 : 509.